Understanding GIS through Sustainable Development Goals

Understanding GIS through Sustainable Development Goals applies a pedagogical shift to learning Geographic Information Science and Systems (herein GIS), as the readers employ the concepts and methodologies on real-world problems. This book provides 16 case studies across most of the Sustainable Development Goals (SDGs) with step-by-step practical instructions using QGIS (Quantum Geographic Information System), an open-source software. It helps readers develop GIS skills on real-world data, while learning the fundamentals including spatial data models, projections, and spatial databases; different cartographic methods, such as graduated symbology, change maps, and dynamic visualization; as well as more intermediate and advanced spatial analysis such as geoprocessing, multiple criteria analysis, and spatial statistics. The topics chosen are taught in secondary and tertiary education institutions, which make this a textbook for all students and educators.

Features:

- Focuses on learning GIS through 16 real-world case studies.
- Introduces an open-source software that can be used beyond the classroom.
- Analyzes Sustainable Development Goals in a global framework and provides an alternative approach to learning GIS.
- Supports both secondary and tertiary educators and improves GIS education at all levels.
- Contains a holistic range of case studies that extend across several disciplines, from geography education, environmental sciences, social sciences, and digital humanities.

Understanding GIS through Sustainable Development Goals
Case Studies with QGIS

Paul Holloway

CRC Press
Taylor & Francis Group
Boca Raton London New York

CRC Press is an imprint of the
Taylor & Francis Group, an **informa** business

Cover Image: A composite image using Shutterstock Image, 17 Sustainable Development Goals set by the United Nations General Assembly, Agenda 2030. Isolated icon set. Stock Vector ID: 2028702848 and background image supplied by Paul Holloway

First edition published 2023
by CRC Press
6000 Broken Sound Parkway NW, Suite 300, Boca Raton, FL 33487-2742

and by CRC Press
4 Park Square, Milton Park, Abingdon, Oxon, OX14 4RN

CRC Press is an imprint of Taylor & Francis Group, LLC

© 2023 Taylor & Francis Group, LLC

Reasonable efforts have been made to publish reliable data and information, but the author and publisher cannot assume responsibility for the validity of all materials or the consequences of their use. The authors and publishers have attempted to trace the copyright holders of all material reproduced in this publication and apologize to copyright holders if permission to publish in this form has not been obtained. If any copyright material has not been acknowledged please write and let us know so we may rectify in any future reprint.

Except as permitted under U.S. Copyright Law, no part of this book may be reprinted, reproduced, transmitted, or utilized in any form by any electronic, mechanical, or other means, now known or hereafter invented, including photocopying, microfilming, and recording, or in any information storage or retrieval system, without written permission from the publishers.

For permission to photocopy or use material electronically from this work, access www.copyright.com or contact the Copyright Clearance Center, Inc. (CCC), 222 Rosewood Drive, Danvers, MA 01923, 978-750-8400. For works that are not available on CCC please contact mpkbookspermissions@tandf.co.uk

Trademark notice: Product or corporate names may be trademarks or registered trademarks and are used only for identification and explanation without intent to infringe.

ISBN: 978-1-032-11573-3 (hbk)
ISBN: 978-1-032-11574-0 (pbk)
ISBN: 978-1-003-22051-0 (ebk)
ISBN: 978-1-032-44647-9 (ebk+)

DOI: 10.1201/9781003220510

Typeset in Times
by codeMantra

Access the Support Material: (https://www.routledge.com/Holloway/p/book/9781032115733 and https://paulholloway.weebly.com/book–understanding-gis-through-sdgs.html.)

Dedication

To Anna, Sofía, and Olivia

Contents

Preface xi
Acknowledgments xiii
Author xv

SECTION I Introductory Material

Chapter 1 Introduction and Context 3

    1.1   Background and Rationale 3
    1.2   This Book 5
    1.3   Structure 6
    1.4   Data and Further Resources 8

Chapter 2 Getting Started with QGIS 9

    2.1   Downloading and Installing QGIS 9
    2.2   Navigating the QGIS Interface 11
            2.2.1   Tabs 11
            2.2.2   Toolbars 12
            2.2.3   Browsing Panel 15
            2.2.4   Plugins 16

SECTION II The Fundamentals of GIS

Chapter 3 Spatial Data Models 21

    3.1   Introduction and Learning Outcomes 21
    3.2   Case Study: SDG13.3 Improving Capacity for Climate Change Mitigation in Ireland 24
    3.3   Case Study Conclusions 39
            3.3.1   Test Yourself 41
    References 41

Chapter 4 Projections 43

    4.1   Introduction and Learning Outcomes 43
    4.2   Case Study: Measuring Fire Size to Support the SDGs 48
    4.3   Case Study: Concluding Remarks 61
            4.3.1   Test Yourself 61
    References 61

Chapter 5 Attributes and Queries 63
- 5.1   Introduction and Learning Outcomes 63
- 5.2   Case Study: SDG9c Targeted Broadband Support Using Attribute Queries 63
- 5.3   Case Study: Concluding Remarks 81
  - 5.3.1   Test Yourself 82
- References 82

Chapter 6 Data Management 83
- 6.1   Introduction and Learning Outcomes 83
- 6.2   Vector Formats 83
- 6.3   Temporary Layers and Broken Links 92
- 6.4   Fixing Geometries 96
- 6.5   Changing Table Fields 100
- 6.6   Managing Data for Success 101
- References 103

SECTION III Cartography

Chapter 7 Location and Thematic Maps 107
- 7.1   Introduction and Learning Outcomes 107
- 7.2   Case Study: SDG14.1 Mapping Plastic Pollution in Coastal Settings 109
- 7.3   Case Study: Concluding Remarks 128
  - 7.3.1   Test Yourself 131
- References 131

Chapter 8 Choropleth Maps 133
- 8.1   Introduction and Learning Outcomes 133
- 8.2   Case Study: SDG4 Visualizing Pandemic School Closures 133
- 8.3   Case Study Concluding Remarks 143
  - 8.3.1   Test Yourself 144
- References 144

Chapter 9 Change Maps 147
- 9.1   Introduction and Learning Outcomes 147
- 9.2   Case Study: SDG15.2 Mapping the Changing Forests of France 147
- 9.3   Case Study Concluding Remarks 161
  - 9.3.1   Test Yourself 162
- References 162

| Contents | ix |

Chapter 10 Dynamic Visualization 163
       10.1   Introduction and Learning Outcomes 163
       10.2   Case Study: SDG15.5 A Year in the Life of a Zebra............ 164
       10.3   Case Study Concluding Remarks 173
               10.3.1   Test Yourself.. 174
       References 174

SECTION IV   Spatial Analysis: Measurements

Chapter 11 Neighborhoods 177
       11.1   Introduction and Learning Outcomes 177
       11.2   Case Study: SDG15.8 Identifying Potential Invasion
              Sites through Neighborhood Analysis.................................. 178
       11.3   Case Study Concluding Remarks 202
               11.3.1   Test Yourself.. 202
       References 203

Chapter 12 Descriptive Statistics 205
       12.1   Introduction and Learning Outcomes 205
       12.2   Case Study: SDG3.6 Analysis of the Spatial Pattern of
              Traffic Accidents in Denver, Colorado, USA 206
       12.3   Case Study Concluding Remarks 220
               12.3.1   Test Yourself.. 221
       References 221

Chapter 13 Density 223
       13.1   Introduction and Learning Outcomes 223
       13.2   Case Study: Modeling Crime Hotspots in Austin, Texas,
              USA, to Support the SDGs 223
       13.3   Case Study Concluding Remarks 233
               13.3.1   Test Yourself.. 235
       References 235

Chapter 14 Interpolation 237
       14.1   Introduction and Learning Outcomes 237
       14.2   Case Study: SDG7 Interpolating Solar Radiation in China ... 238
       14.3   Case Study Concluding Remarks 246
               14.3.1   Test Yourself.. 247
       References 247

SECTION V Spatial Analysis: Geoprocessing

Chapter 15 Site Selection – Multiple Criteria Assessment 251

    15.1   Introduction and Learning Outcomes 251
    15.2   Case Study: SDG6 Locating a New Wastewater
           Treatment Plant in Uganda 252
    15.3   Case Study Concluding Remarks 264
           15.3.1   Test Yourself 265
    References 265

Chapter 16 Risk Analysis – Unique Condition Unit 267

    16.1   Introduction and Learning Outcomes 267
    16.2   Case Study: SDG11.5 Landslide Risk in Arequipa, Peru 268
    16.3   Case Study Concluding Remarks 284
           16.3.1   Test Yourself 285
    References 285

Chapter 17 Site Selection – Map Algebra 287

    17.1   Introduction and Learning Outcomes 287
    17.2   Case Study: Locating a New Electric Vehicle Charging
           Station in Vancouver, British Columbia, Canada 289
    17.3   Case Study Concluding Remarks 301
           17.3.1   Test Yourself 301
    Reference 301

Chapter 18 Route Selection 303

    18.1   Introduction and Learning Outcomes 303
    18.2   Case Study: SDG11.2 What is the Best Route to
           Take to the Train Station? 304
    18.3   Case Study Concluding Remarks 314
           18.3.1   Test Yourself 314
    References 314

Epilogue 317

Index 321

Preface

This book explains and demonstrates various theoretical and practical components of Geographic Information Science and Systems (herein GIS). It is a textbook with practical examples in QGIS (Quantum Geographic Information Systems), intended for readers who want to familiarize themselves with the field. The book does not aim to cover the whole field, but rather explains the main steps in understanding and implementing real-world practical examples that span topics including the fundamentals of GIS (e.g., spatial data models, projections, spatial databases), best cartographic practice, as well as more intermediate and advanced spatial analysis (e.g., geoprocessing, multiple criteria analysis, spatial statistics). It should therefore prove useful to lower- and upper-level undergraduate and graduate students taking courses in GIS, across different disciplines. It should also be of relevance to secondary school educators who are looking to develop material as part of new requirements of geography curriculums.

The book implements a case study approach to learning GIS that aligns with the U.N. Sustainable Development Goals (SDGs). These are comprised of 17 goals that address global challenges to poverty, inequality, climate change, environmental degradation, and peace and justice. Within this book, there are 16 case study chapters that span at least ten of the SDGs, with examples drawing from several open-source databases to provide a global perspective to the practical elements, including climate change, litter, ecology, geology, energy, criminology, transport, and urban planning. Subsequently, this book links GIS theory and practical work with on-the-ground global policy that is supporting the development of a more sustainable future.

Readers should start with Chapter 1, which explains in more detail what the book is about, as well as the rationale for its content. The book advances incrementally from Chapters 1 to 18, with the content in latter chapters supported by the fundamental components introduced in earlier chapters, although it is possible to complete the chapters in isolation.

Acknowledgments

I am extremely grateful to James O'Mahony for reading, testing, re-testing(!), and commenting on the entire manuscript. I am also greatly appreciative of the considerable efforts of several others who reviewed significant portions of the book at various stages, offering excellent feedback, suggestions, and corrections, including Darius Bartlett, Siddharth Joshi, Mike Murphy, and Dipto Sarkar. I would also like to thank TDF for the inspiration for Chapter 13. I would like to thank the students who classroom tested this book, often at much earlier drafts, providing very useful feedback to improve the case studies presented here.

I thank my Editor at CRC Press/Taylor & Francis Group, Irma Britton, and Editorial Assistants, Chelsea Reeves and Shannon Welch, for their support and assistance.

The writing of this book was supported by the approval of research sabbatical leave from the College Sabbatical Research Leave Committee at University College Cork.

This book builds upon the principle of open software and data. Therefore, I would like to thank all the individuals, organizations, and contributors who have licensed data and software in the public domain, allowing such a diverse and geographically widespread range of case studies to be covered.

Finally, I thank my family, particularly Anna, for her support in everything I do, but particularly the marathon that has been this textbook. Your unconditional support and love mean everything and have made the long nights over the last 12 months much more bearable.

Author

Paul Holloway is a lecturer in Geographic Information Science and Systems in the Department of Geography at University College Cork, a principal investigator in the Environmental Research Institute at University College Cork, and the Vice President of the Irish Organization of Geographic Information. Paul's teaching and research interests include using GIS and spatial analysis to address a suite of environmental, ecological, and geographic issues. He teaches several undergraduate and graduate GIS courses at UCC and was presented with the President's Award for Excellence in Teaching in 2019. Paul's research is widely published, with more than 35 peer-reviewed publications in GIS-focused journals, and he is currently involved in several nationally and internationally funded projects exploring the application and development of GIS across the biodiversity, agriculture, energy, health, and transport sectors.

Section I

Introductory Material

1 Introduction and Context

1.1 BACKGROUND AND RATIONALE

Geographic Information Systems and Science (herein GIS) and related spatial technologies have evolved substantially in recent years. Geographic Information Systems are defined as the technology available for the acquisition, management, and processing of spatial data, while Geographic Information Science is the study of the fundamental issues, theories, and concepts behind the technology. The definition of GIS will change depending on who you talk to, but across definitions, there is a consistent set of components that allow the storage, management, and analysis of spatial data. GIS in this book is intended to be all encompassing, and through the next 17 chapters, we explore practical skills and challenges associated with both the Systems and the Science of Geographic Information.

Much of our daily lives are now underpinned by spatial data and GIS. If we think about the technology that we hold at our fingertips in the form of smartphones, laptops, and personal computers, many of the search algorithms or route-finding apps are underlain with maps containing spatial data. We can use this technology to find hotels, restaurants, and even romantic partners within a geographic distance of our current location, querying this selection based on a set of attribute data related to these features, such as hotel star ratings. There are literally thousands of real-world applications of GIS that span an array of disciplines. However, with a wider uptake of GIS and related spatial technologies, there persists the need to ensure an understanding of the topic to ensure best practice and robust interpretation of the processes and patterns we explore through these systems.

GIS users and specialists are tasked with important questions, including: How should geographic data be compiled and measured? How do people perceive real-world geographic variation? How do we collect spatial data? How do users handle uncertainty or inaccuracies in spatial data? How can we develop effective maps and visualizations of our spatial data and processes? Which spatial analytical technique should be used to measure distance, density, or geostatistics? What is the best method to overlay layers of spatial data within GIS? GIS users therefore have the gargantuan task of finding a way to represent our infinitely complex and dynamic world to best support spatial cognition and reasoning.

We are now seeing a widespread uptake of GIS in secondary education and a sustained engagement across a variety of disciplines in tertiary education. For example, my current GIS courses at University College Cork, Ireland, are core and optional requirements for geographers, archaeologists, geologists, environmental scientists, zoologists, biologists, and digital humanitarians. The role of GIS within the education sector has subsequently developed drastically over the last 20–30 years, such that we are beginning to see more emphasis placed on the real-world applicability of GIS along with material that keeps pace with the continued advancements in technology. Therefore, there remains demand for resources that cover the fundamental (theoretical and conceptual) issues of GIS, but with an added focus on the applied nature of GIS. This is particularly relevant in tertiary education where GIS is beginning to be taught in disciplines outside of geography, or in secondary institutions where GIS is being taught without the resources available to staff who may never have been explicitly trained in such a subject.

Developing a case study approach with an international perspective to understanding GIS within the human and environmental realms requires a universally recognized framework. Sustainability challenges are well recognized in international policy agendas, notably in the 2030 Agenda for Sustainable Development, which sets out the sustainable development goals (SDGs). The United Nations (UN) SDGs are comprised of 17 goals that address global challenges related to poverty, inequality, climate change, environmental degradation, peace, and justice. Subsequently, the SDGs are discipline independent, with a wide-reaching impact for many of the interdisciplinary challenges society is currently tackling. For each goal, several targets have been defined, and for (most of) the targets, there are several indicators and metrics for measuring progress toward those targets and, ultimately, the overall goal.

For example, SDG3.6 aims to halve the number of global deaths and injuries from road traffic accidents. By knowing where car accidents are occurring, and then exploring the patterns associated with serious injuries and fatalities, mitigation and preventative measures can be directed to areas that could support successful attainment of this target. We explore how GIS can support this target in Chapter 12. The SDGs therefore provide a holistic overview to understanding GIS that spans disciplines, but also fosters the development of action-orientated global citizens who will support efforts for a sustainable future (Figure 1.1).

FIGURE 1.1 The 17 United Nations sustainable development goals. (Shutterstock: Stock Vector ID 2028702848.)

The transformative vision for the SDGs is that the needs of the most vulnerable are met and no-one is left behind. Subsequently, the big societal and environmental issues that the SDGs tackle can cover some very sensitive topics. I'm conscious when considering chapters that touch upon crime, perceptions of safety, traffic accidents, and natural disasters that it can become easy to treat the data as simply data, and not as actual events that have impacted individuals. Every crime, accident, fatality, or other event that we explore in this book did happen. Therefore, while the focus herein is to demonstrate how GIS can be used to support SDGs, the case studies are applications of possible questions that can be answered. They are by no means intended to trivialize the experiences of anyone who has undergone these or equivalent events. Moreover, in many of the case studies, we only use a subset of possible variables to demonstrate the tools. In reality, everyone's viewpoints will differ, and the choice of different factors and variables is again not intended to be all-knowing, but rather provide a simplified example to demonstrate the GIS tools. As is the aim of any educator, I hope that such ideas resonate with you for your own research and work, and that these methods can be used by you, the future experts, to tackle these pressing environmental and societal issues.

There is also a growing momentum around open-source GIS tools such as QGIS (Quantum Geographic Information Systems), with the user-base continually growing and an increasingly engaged community. QGIS is a free and open-source software that supports the storage, management, presentation, and analysis of spatial data. It is platform independent, meaning it works across a range of operating systems (OS) including Windows, Mac OSX, Linux, Unix, and Android. QGIS has a range of in-built core tools and functions that allow users to work with this spatial data, as well as supporting plugins which provide access to community developed tools. There are a variety of plugins that range from experimental to fully developed and seamlessly integrated into the QGIS functionality. QGIS has several dependencies that include GEOS and SQLite, and is primarily coded using C++. Other dependencies for specialized tools include GRASS GIS, GDAL, and PostGIS. Subsequently, there is a lot of functionality available within the QGIS interface. This book focuses on the core functions and plugins, as the objective of this book is to support the understanding of GIS.

1.2 THIS BOOK

This book subsequently has the following interlinked objectives: (1) to provide an overview of GIS, whereby the concepts and methodologies are explained, demonstrated, and tested using real-world case studies; (2) to provide support to instructors and students involved in GIS courses and researchers and/or scientists interested in applying GIS to their problem-set; (3) to demonstrate real-world applications through the prism of SDGs; and (4) to develop case studies in QGIS, which will provide an open-source software environment to learn GIS. As such, this book provides an alternative approach to learning GIS; rather than being theory-led, it applies the theory through detailed practical instructions.

The book covers various aspects of GIS to provide an up-to-date resource. Step-by-step practical instructions allow readers to use case studies and real-world data to develop their GIS skills, while learning the fundamentals of GIS theory. This book covers topics including the fundamentals of GIS (e.g., spatial data models, projections, and spatial databases), best cartographic practice, and more intermediate and advanced spatial analysis (e.g., geoprocessing, multiple criteria analysis, and spatial statistics). Topics have been chosen that align with key content taught in education institutions and the learning outcomes of those curriculums.

The material in this book is of relevance to lower- and upper-level undergraduate and graduate students taking courses in GIS, across different disciplines. It should also be of relevance to secondary school educators who are looking to develop material as part of new requirements of geography curriculums. Finally, it is also useful to researchers and scientists interested in learning techniques and technologies associated with GIS, specifically the open-source software QGIS.

This book assumes no prior knowledge of GIS theory or QGIS. The case studies are drawn from several open-source databases to provide a global perspective to the practical elements. Real-world case studies related to climate change, litter, ecology, geology, energy, criminology, transport, and planning will be applied as we work through this book. This book provides readers with a specific focus on learning GIS through case studies, rather than learning the theory and then application of practical work. The implementation through open-source software provides a knowledge base that can be freely used by individuals beyond any institutions they are currently situated in. The focus on SDGs situates the case studies in a prominent international framework that is currently shaping policy, which will support employment opportunities within the GIS, information and communications technology (ICT), and thematic sectors.

1.3 STRUCTURE

This book contains five sections:

- Introductory Material
- The Fundamentals of GIS
- Cartography
- Spatial Analysis: Measurements
- Spatial Analysis: Geoprocessing

The first section of the book introduces the material needed to progress through the subsequent chapters. The rationale behind the motivations for this book has just been provided as well as an overview of the SDGs. In Chapter 2, an overview of the QGIS software is provided, as well as information on how to download and install the software and a brief introduction to the user interface.

The second section of this book deals with the fundamentals of GIS. For GIS to address geographic questions, it must have a way of storing quantitative data of different features alongside spatial information. GIS essentially generates a layered view of the world, with data stored in separate 'layers', which we build up to generate complex models of the real world. However, to achieve this we need to layer the foundations to successfully represent this infinitely complex world within a computerized system. The basic elements of spatial or geographic data include the location of such features (i.e., coordinates), the attribute data (i.e., the phenomena of interest), and a transformation to convert geographic locations into a position on a flat surface (i.e., a projection). I term these the fundamentals of GIS as these are in my opinion the building blocks that are needed to truly understand the more advanced geoprocessing and measurement operations available to us within GIS.

The third section of this book deals with cartography. GIS has supported the development of cartography through the availability of easy-to-use tools that can create professional maps. As a (partial) result of this, maps are one of the predominant methods through which we, GIS users/specialists, communicate our data, information, and findings, which in turn supports knowledge generation. Maps enable us to facilitate a spatial understanding, allowing us to store information, reveal patterns, and illustrate relationships. In short, they provide a medium to support us in our goal of understanding our infinitely complex world. Importantly, they provide an aesthetic and powerful representation of the questions we are attempting to answer. The adage 'a picture is worth a thousand words' is apt, particularly if we consider replacing the word 'picture' with 'map'. This method of visual communication consolidates verbal and written communication, and as such, the ability to create professional maps is a necessary and highly valuable GIS skill. Moreover, cartography is an essential contributor to achieve the SDGs, as without the presentation and visualization of work, the demonstration of targets and information can be overlooked.

While the process of generating maps has become simpler with the advent of GIS, cartographic principles remain important, and a gap in the knowledge of cartography can lead to poorly designed and ineffective maps. Therefore, it is the aim of the third section to provide an overview of cartography, considering different map types, before looking beyond static maps and using GIS to generate dynamic outputs. Throughout these chapters, we explore the basic characteristics of maps, including map balance, the process of abstraction, symbology, and the overall art of map-making. Entire careers and books have been dedicated toward cartography, and as such, this section is not to be considered a comprehensive guide to cartography, but rather a starting point with which to take the geographic data and present it as clearly and effectively as possible.

The fourth and fifth sections of the book develop our spatial analysis. We actually begin the process of spatial analysis in the earlier chapters, but up to this point, the spatial analysis is largely exploratory in nature. In the fourth section, we undertake more advanced analytical methods to generate new information that can inform decision-making. This section demonstrates how GIS quantifies and measures spatial data, creating new geographic information, such as neighborhoods, spatial statistics, density surfaces, and interpolations. To undertake rigorous and robust spatial analysis, we must have a strong grasp of the building blocks of the system, i.e., the fundamentals. Therefore, we should continue to critically think about our decisions related to spatial data models, projections, and the attribute table, and these concepts will continually be reinforced.

In the fifth and final section of this book, we explore various methods of implementing multiple criteria assessments (MCAs) across data models. MCA integrates and combines multiple datasets through geoprocessing, which are then used to identify (or assess) areas that satisfy a pre-specified range of criteria. We consider four examples of MCA utilizing several tools that we have at our disposal, working across both vector and raster formats, posing the following questions:

- 'Where is the most suitable location to build a new wastewater treatment plant in Uganda?' to support SDG6.1.
- 'Where are the most likely locations for landslides in Arequipa, Peru?', supporting SDG11.5.
- 'Where is the most suitable site to support a new electric vehicle charging station in Vancouver, Canada?', supporting SDG7.1 and SDG13.2.
- 'What is the best route to take between tourist sites and the train station in Edinburgh, Scotland?', specifically supporting SDG11.2 and SDG5b.

1.4 DATA AND FURTHER RESOURCES

All the data used in the case studies in this book have been sourced from existing open data portals and/or geoportals. Data is available through the Supplemental Material of the book at the following location: https://www.routledge.com/Holloway/p/book/9781032115733. Data is also available at: https://paulholloway.weebly.com/book-understanding-gis-through-sdgs.html.

All data has been stored in zipped folders, which need to be downloaded, extracted, and saved to your local device. If you do not extract the data, it will not be compatible with QGIS and you will not be able to use it within the software.

The author will make periodic updates to the website as needed. For example, new Long-Term Releases of QGIS (every year in February) may result in changes to tools, functions, or plugins. Please refer to this website if you have any queries as you work through the book.

2 Getting Started with QGIS

2.1 DOWNLOADING AND INSTALLING QGIS

1. Navigate to the QGIS website: https://qgis.org/en/site/
    Here is the QGIS homepage detailing current community news, updates, and download instructions across the different platforms.
2. Scroll down the page to the 'Download Now' button.
    There are two versions of QGIS that we can download. This represents the current version and the long-term release (LTR). The current version represents the most up-to-date version with the most recently developed features, while the LTR provides a common platform with which to build training material as it represents a standardized version that will not change for a year. Therefore, the rest of this book will subsequently follow the LTR at the time of writing, which is 3.22.5 LTR.
3. Click the 'Download Now' button.
    This takes us to an interface outlining download instructions for Windows, Mac OSX, Linux, Unix, and Android. QGIS is software independent, meaning we can install this on a range of operating systems, including those just mentioned. The following instructions within the book are for a Windows device, but the key components should be broadly similar for all operating systems. Instructions for some other operating systems are provided on the QGIS website, as well as this book's website that was provided in the previous chapter. For Windows, there are two options, OSGeo4W and Standalone installer. We are interested in downloading the OSGeo4W Network Installer, as this contains the LTR.
4. Click on the Installer to download the setup.exe file.
5. Click on the setup installer that has just been downloaded to the device we are working on.
6. We are given two options, Express Install and Advanced Install. We want to select the Advanced Install, and then click next.
7. If new to QGIS, select the download source 'Install from the Internet' and then click Next.
    Finally, we are asked to Select Root Install Directory. I suggest keeping this as the default, as this will ensure you do not run into any complications further down the road with regard to naming conventions.
8. Keep Select Root Install Directory as default and click Next.

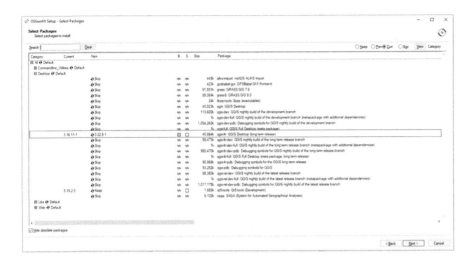

FIGURE 2.1 Screenshot of where to download and install the long-term release desktop version of QGIS in order to follow the instructions provided within this book.

9. Following this, we need to Select Local Package Directory. Again, unless you have any strong desires to change the name or specify somewhere else, I recommend just keeping this as the default option before clicking Next.

The next page requires us to choose the method of Internet Connection. In most instances, the default option of 'Direct Connection' should suffice; however, if working on a machine that has a proxy you will need to change this and specify the details as requested.

10. Keep the default settings and click Next.
11. In the next screen, we are asked to Choose a Download Site, and again unless directed otherwise I would select 'http://download.osgeo.org'. Click on it once, and then choose Next.

Here we are taken to an interface to 'Select Packages', as shown in Figure 2.1. This is where we choose the LTR. If we expand the packages within the Desktop Category by pushing the + button, a long list of packages appears (note you may need to scroll across to the right to get the package names or increase the size of the dialogue box).

12. Scroll down to where we can see the LTR and click on 'Skip'. 'Skip' changes to '3.22.5.1'. This is the download option we want, so click Next.

The previous LTR was already installed on my PC, hence that is why 3.16.11.1 is noted next to version 3.22.5.1.

13. There may be some dependent packages that require downloading. It is recommended to install these as well. Ensure the box is ticked and click Next.

QGIS is now downloading. This could take a little while, depending on machine and Internet Connection, so take a pause here before moving onto the next sub-section that will introduce the QGIS interface. For reference, the last time I downloaded this, it took approximately one hour, so plenty of time for a coffee or tea.

Getting Started with QGIS

2.2 NAVIGATING THE QGIS INTERFACE

14. Navigate to the download location and open QGIS.

    The QGIS interface will be our primary home as we progress through this book, so we briefly introduce some techniques for using this software. As with any software, it is always good practice to get into the habit of saving regularly. The home page has various information regarding previous projects and news items. As this is most likely the first time we are using QGIS, the Recent Projects should be Empty.

15. To get started, double click on the Project Template.

    This should generate a white page that can be populated with spatial data. This is known as the map canvas in QGIS. It is one of the most important components of the software as it allows us to interact with our spatial layers. There are also a range of toolbars, panels, and options. An overview is provided in Figure 2.2.

    There are a lot of features associated with QGIS, too many to document here. As such, the different features, tools, and buttons will be introduced periodically as we progress through the book. However, some key components of the interface are outlined in Figure 2.2, and more information is provided below.

2.2.1 Tabs

There are multiple tabs that contain numerous commands for working with the spatial data. Below is a quick introduction to them:

Project – The project tab manages our projects, providing the opportunity to save, open existing projects, as well as opening the print layout commands. It also controls the different licenses that are available.

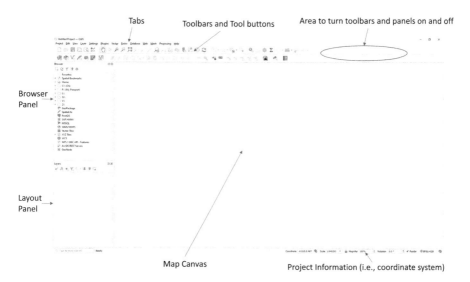

FIGURE 2.2 Overview of the QGIS interface.

Edit – The edit tab includes undo or redo commands, as well as copy and paste information and add shapes and edit features. In short, it provides support to manage, change, and transform our data, which is how we build up a layered representation of the world.

View – The view tab provides us with the options to interact with the map interface, such as zooming in or out, and managing which layers are visible. It also provides more functionality, including toggling between elements and access to some standard tools such as measurement.

Layer – The layer tab allows us to manage the specific layers (or features) that we have in our map. It allows us to add data, edit layers, as well as setting coordinate information and removing information.

Settings – The settings tab provides us with a range of specific user options and keyboard shortcuts to manage our QGIS experience.

Plugins – The plugins tab allows us to manage and install plugins, which is essential for benefiting from bespoke tools that are not part of the QGIS core functionality. We will add a plugin later in this chapter.

Vector – The vector tab is where we find the geoprocessing tools, models, and scripts to work with vector data.

Raster – The raster tab is where we find the geoprocessing tools, models, and scripts to work with raster data.

Database – The database tab is where we find the database manager, which will allow us to manage our spatial datasets.

Web – The Web tab allows us to search for QGIS support and services.

Mesh – The Mesh tab allows us to work with data that are stored in specific mesh format.

Processing – The processing tab provides us with useful information regarding the processing history and results that we undertake.

Help – The help tab provides useful features to support our QGIS experience.

As we progress through this book, we utilize several of these tabs.

2.2.2 Toolbars

There are multiple toolbars with numerous tools that can be used by clicking on them. When we first install QGIS, there should be several that are preloaded, including the following:

Project Toolbar – This toolbar provides access to buttons that allow us to manage our QGIS projects, including opening new projects and saving them.

Map Navigation Toolbar – This provides access to buttons that allow us to navigate the map canvas, including zoom and pan.

Selection Toolbar – This toolbar provides access to buttons that allow us to select different spatial and attribute features.

Attributes Toolbar – This toolbar provides access to buttons that allow us to identify attribute values, measure distances, and open the processing toolbox.

Data Source Management Toolbar – This toolbar provides access to buttons that allow us to connect with existing data sources and create new data.

Digitizing Toolbar – This toolbar provides access to buttons that allow us to edit our features, including the spatial and attribute information.
Label Toolbar – This toolbar provides access to buttons that provide us with the ability to apply labels to the data.
Help Toolbar – This provides access to the button that opens the in-built QGIS help functions.
Plugins Toolbar – This provides access to all buttons that we add through the Plugin Tab.
Web Toolbar – This provides access to a button that enables meta data searches.

The toolbars that are included in the default interface are those that are used widely and regularly by most GIS users, but there are in fact several more. However, having all these open all the time can lead to a cluttered and messy interface. Therefore, many are hidden. We can activate these by simply turning them on. If we right click in the gray area highlighted in Figure 2.2, it will open a list of all possible toolbars. Please note that the color of this area can change depending on the OS and personal device settings. We can then simply turn these toolbars on and off, depending on which tools we want visible or need for our GIS work.

16. Click on the gray area to switch toolbars on and off.
17. Untick Label Toolbar.

This should remove this toolbar from the Map canvas. We do not use this toolbar during this book, so we can keep this switched off.

18. Click on the gray area again, and tick Processing Toolbox.

This should open a new toolbox that either docks on the right of the screen or floats, shown in Figure 2.3. This toolbox contains all possible processing tools available within QGIS and dependencies. As we progress through this book, we use a combination of tools that are found in the tabs, toolbars, and processing toolbox, and we explore the different methods of accessing these.

There are obviously too many tools to mention at this point, but it is important to try and learn what these tools are called and what they do as we advance through our GIS learning experience. These tools all have names and every tool that is used in this book is detailed with the visual graphic, its name, and the location it is found (i.e., Data Management Toolbar) in the Glossary that is located as part of the supplementary material for this book online.

Having taught GIS using QGIS over the past few years, there have been instances where buttons have not been available or have been grayed out. The process of turning the toolbar off and then on again and then restarting QGIS has generally resolved this issue, or if the tool is affiliated with a plugin (detailed in Section 2.2.4) the same process of reinstalling the plugin works in a similar fashion. Therefore, if you find yourself in a similar position of a tool being 'grayed out', try turning off the relevant toolbar before turning it back on again.

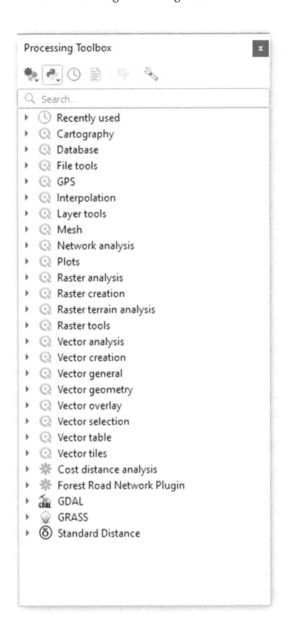

FIGURE 2.3   Screenshot of the processing toolbox.

Getting Started with QGIS 15

2.2.3 Browsing Panel

The Browser Panel should already be open in the QGIS homepage and is labeled in Figure 2.2. If it is not visible, then we can turn it on using the method we have just implemented.

19. On the gray area where the toolbars are located, right click, and ensure that the Browser Toolbar is selected.

    The QGIS Browser is an in-built panel and a standalone application that allows us to navigate our filesystem and manage our spatial data. Here we have access to our spatial data and our files so that we can efficiently manage them in a spatially referenced database. We return to the Browser with a lot more detail in Chapter 6.

    There are also various options that link to different basemaps, spatial databases, and dependencies.

20. In the Browser Panel, expand the XYZ Tiles. There should be one option: OpenStreetMap.
21. Double click on the OpenStreetMap.

    This should have populated the map with a reference map that contains OpenStreetMap features, shown in Figure 2.4. We use this reference map or basemap regularly throughout this book, as such a feature allows us as GIS users to reference ourselves on the globe, which can be very useful to ensure that our spatial data is located in the correct place. It should have also added the layer OpenStreetMap to the Layers Panel. We introduce the Layers Panel in much more detail in the next chapter.

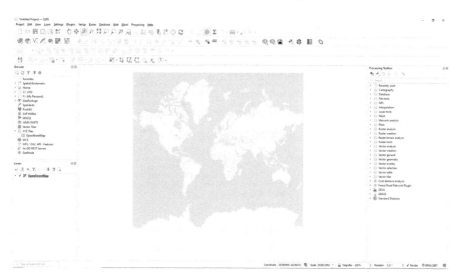

FIGURE 2.4 Screenshot of the QGIS interface with the OpenStreetMap basemap visible in the map canvas. Basemap is the OpenStreetMap XYZ tiles which is © OpenStreetMap contributors and available under the Open Database License. Please see https://www.openstreetmap.org/copyright.

22. Click on the Zoom-in tool to explore the map, or use the mouse to scroll and pan around the map.

2.2.4 PLUGINS

Plugins are developed independently of the QGIS organization, meaning they are created by independent researchers and organizations to provide additional functionality and tools. Therefore, we have several additional tools that can be used to manage, store, and analyze our spatial data. We have visualized the OpenStreetMap in-built basemaps that can be used for spatial reference in the previous section; however, there are several more basemaps that are available through a plugin QuickMapServices. We will install this plugin now and demonstrate how we can install additional basemaps.

23. Navigate through the tab Plugins > Select Manage and Install Plugins.

    This will open a dialogue box, shown in Figure 2.5, which contains all available plugins. The default is to show all plugins that are currently installed. As just stated, there are some plugins that are in-built.
24. Click on the Not Installed tab. Here we can see the long list of available plugins that could be used.
25. Click on Settings and tick 'Show also experimental plugins', this further increases the list of available plugins that could be installed.

    Experimental plugins are usually in the early stages of development and should be considered as proof of concept or beta tools, rather than for

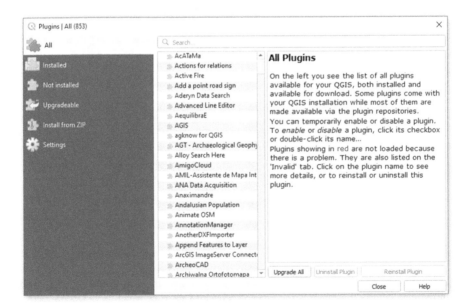

FIGURE 2.5   Screenshot of the QGIS plugin interface.

Getting Started with QGIS   17

FIGURE 2.6 Screenshot of the process to insert the OSM Standard Basemap using the QuickMapServices plugin in QGIS.

widespread use. Therefore, this book will focus on plugins that have been tested and fully integrated.

26. Re-click on the Not Installed tab, and in the search tab, type QuickMapServices.
27. Select this option. A description of the plugin is provided, including any metadata of the tool.
28. Click Install Plugin.
29. Click close.
    There should now be some new options in the Plugin Toolbar.
30. Click on the WebMapServices button.
31. There should be two in-built basemaps, NASA and OSM.
    Select the OSM Standard basemap, as shown in Figure 2.6.

The features portrayed here match exactly those that were present in the earlier basemap. We can connect to different basemaps and will build upon this as we progress through the book. In the next chapter we start to populate the project with spatial data and begin the process of understanding GIS through SDG case studies.

Section II

The Fundamentals of GIS

3 Spatial Data Models

3.1 INTRODUCTION AND LEARNING OUTCOMES

In this chapter, we delve into how to represent our infinitely complex world within a computerized system. Our job as GIS users is to determine the most suitable conceptualization and methodology with which to represent such phenomena. We cannot just put spatial data into a computer and expect the technology to intuitively make sense of it. Instead, we must reduce the complexity of the real world to manageable proportions. To do this, we use spatial data models.

There is a saying in GIS that approximately 80% of data has a spatial component. While this figure is somewhat arbitrary and has been the subject of research and debate over the years (Hahmann & Burghardt 2013), the premise holds up. Most geographic features, phenomena, or events happen somewhere. In other words, they do not occur in a void. Therefore, if we know where something occurs, we can add a spatial dimension to our data. This provides us information on the geographic attributes of such data, creating spatial data. Spatial data can include a broad range of data types, including demographic data (e.g., births, deaths, illness, and age), socio-economic data (e.g., race, ethnicity, income, and education attainment), environmental data (e.g., soils, rocks, rivers, trees, and organisms), and climate data (e.g., temperature and precipitation).

The basic elements of spatial or geographic data include the location of such features (i.e., coordinates), the attribute data (i.e., the phenomena of interest), and a transformation to convert geographic locations into a position on a flat surface (i.e., a projection). We return to projections in Chapter 4 and queries of the spatial and attribute data in Chapter 5.

For this chapter, it is sufficient to understand the basic components of coordinate systems. Coordinate systems are simply a method for storing the geographic location of features, consisting of three basic elements, shown in Figure 3.1:

- An origin – which is then the location to which every other point can be referenced to
- A defined direction – i.e., east, west, north, or south
- A distance in that direction – specified in the units of interest

All spatial data must have these elements, regardless of the model that is used. A model on the other hand is an abstract representation of a system or process. This can be as simple as a model airplane, model village, or a statistical model. Through the process of abstraction, models allow us to represent very complex processes in a manageable format, which aids in defining problems more precisely. For example, maps are geographic models. These cartographic representations simplify very detailed and extensive landscapes to more manageable scales. Maps have been the

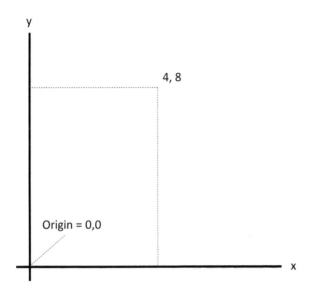

FIGURE 3.1 Example of a coordinate system, with the three basic elements including the origin, direction, and distance.

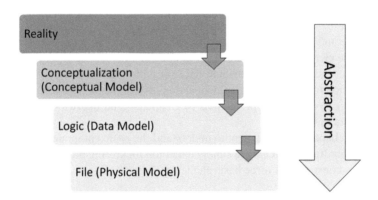

FIGURE 3.2 The process of increasing abstraction for representing spatial data in GIS.

predominant format of conveying spatial data for centuries, providing information on the physical and social geography of different areas. In GIS, we simply replace the map with a database, and the database becomes the medium through which reality is represented.

Subsequently, a spatial data model is the method through which we increase the abstraction from the human-orientated toward the computer-orientated, all while maintaining the geographic information associated with our phenomena. To achieve this, we need to define a conceptual model of space and then a logical or data model for implementing this in a physical model (i.e., digital file), conceptualized in Figure 3.2.

Spatial Data Models

In GIS, there are two predominant conceptualizations of space: discrete objects and continuous fields. Discrete objects have well-defined boundaries and occupy the space in which they occur. We can often think of this space as an empty world, underlain of course by a coordinate system, but waiting to be populated by such entities. Continuous fields on the other hand conceptualize variables as continuous in space, with the geographic world represented by a finite number of variables, each with a value defined at every possible position. These represent the two predominant conceptual models, which subsequently need to be coded into a data model.

The data model determines how we represent the data in our spatial database. There are two basic families of logical data models in GIS: the object-focused approach and the field-orientated (or tesseral) approach. The object-focused approach is based on a recognition of discrete objects and/or entities and generally uses spatial primitives of points, lines, and polygons (i.e., areas) to represent these features. The field-orientated approach generally uses grids or tiles (usually square, but can be rectangular, triangular, or hexagonal) as the basic spatial primitive. The object-focused approach is usually represented by vector GIS, while the field-orientated approach is usually represented by raster GIS; however, it should be noted that these are not exclusive categorizations, as we can see as we move through this chapter.

In the vector data model, discrete locations are represented as points, all lines are subsequently captured by points, connected precisely by straight lines, while a polygon is represented as a series of points or vertices connected by straight lines, consisting of three (or more) straight sides. The raster data model divides the space into an array of cells, assigning attributes to each one of the cells. All geographic variation is expressed by assigning values to these cells, across the entire study area.

Spatial data models therefore provide the building blocks to view our world within a GIS. Moreover, they allow prediction of the underlying processes, so we can estimate how changes to those systems may alter the outcomes, which ultimately promotes understanding of the geographic systems we study. Models also permit the impossible and allow experiments that may not otherwise be possible. Central to the SDGs is the mitigation and adaptation to climate change. Through simplifying climatic processes into models (albeit complex models), we can begin to explore which locations may be exposed to extreme climate conditions in the future, allowing more informed decision making with regard to institutional and individual action.

SDG13 proposes urgent action to combat climate change and its impacts, with target 13.3 to improve education, awareness raising, and human and institutional capacity on climate change mitigation, adaptation, impact reduction, and early warning. There are obviously several ways in which such information can be integrated within educational programs, but a simple and effective method is to use the layering of GIS to explore climate projections alongside education institutes. This chapter will subsequently focus on how environmental features (i.e., temperature and precipitation) and infrastructure (i.e., schools and roads) are conceptualized and modeled with the overarching SDG aim to identify the impact of climate change on infrastructure and

FIGURE 3.3 Screenshot of the Roads – OSi National 250k Map of Ireland data.

people, specifically in Ireland. By the end of this chapter, you will have completed four learning outcomes, and you should be able to:

- Critically discuss geographic information and the use of spatial data models to represent geographic objects and phenomena
- Navigate around the QGIS interface using a variety of basic tools and functions
- Perform basic spatial operations by overlaying different spatial data models
- Convert between spatial data models

3.2 CASE STUDY: SDG13.3 IMPROVING CAPACITY FOR CLIMATE CHANGE MITIGATION IN IRELAND

The data for this chapter consists of a shapefile for counties in Ireland, obtained from the Central Statistics Office (CSO 2013) licensed under Creative Commons Attribution 4.0. The attribute table has been reduced to contain fewer attributes. A shapefile representing post-primary schools in Ireland published by the All-Island

Spatial Data Models

Research Observatory (AIRO 2016) licensed under Creative Commons Attribution 4.0. Another shapefile representing the Ordnance Survey Ireland (OSi) National 250k Map of Ireland road network (OSi 2016) also licensed under Creative Commons Attribution 4.0. Finally, there are three raster layers representing current and future climate from Hijmans et al. (2005), licensed under Creative Commons Attribution-Sharealike 4.0 International License.

Firstly, we want to open a New Empty Project in QGIS:

1. Open QGIS.
2. Start a New Empty Project.
3. If not completed already, download the zipped folder from the website labeled 'Ch3_Data.zip' and extract the folder.

We should have an empty map. Time to start populating it with some data...

4. Navigate through the tab: Layer > Add Layer > Add Vector Layer.
5. In the source tab, navigate to where you have extracted the folder using the ... button.

There are several layers saved within the folder. We want to add Roads; however, there are various files with such a name. A shapefile is a file format for storing the geographic location and attribute information of geospatial features. Shapefiles can be points, lines, and polygons, and a very common method of storing vector data. There are actually four files that make up the shapefile. The three mandatory files are *.shp, *.shx, and *.dbf, and the *.prj is the projection file that specifies which coordinate system defines the spatial units. The term 'shapefile' relates to the *.shp file, but this file alone is insufficient for the file to be read in a GIS. Therefore, it is important that when saving or transferring shapefiles we move all the files associated with it. However, QGIS allows us to enter this file and will link to the associated files.

6. Double click on the *.shp file for Roads, and add this to the project.

The project has been populated with a layer that depicts the roads at a national scale in Ireland, as shown in Figure 3.3. This should also be present in the Layers Panel now.

7. Repeat this process for the other two layers in the folder: Counties and Schools.

Your screen should resemble Figure 3.4, with three layers in the Layers Panel, and the counties and schools clearly identified. However, we can no longer see the roads layer. The order that layers appear in the Layers Panel are the order in which they are drawn in the map canvas. We have two options, either change the symbology of the counties, or reorder the layers. We are going to reorder the layers.

8. Click and hold the layer Roads in the Layers Panel.
9. Drag this above the Counties layer, and release the mouse.

FIGURE 3.4 Screenshot of all counties, schools, and roads loaded in QGIS; however, note that the roads layer is currently hidden.

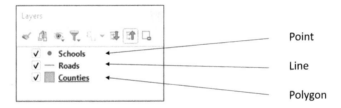

FIGURE 3.5 Layers panel overview of points, lines, and polygons.

Roads is now visible in the map canvas. The Layers Panel should now resemble Figure 3.5:

In vector GIS, real-world objects can be represented as points, lines, or polygons. The term features is often used to refer to real-world objects; features have a location, shape, and symbol. For example:

- Schools and cities can all be represented as points.
- Roads and rivers can all be represented as lines.
- Counties, fields, and parks can all be represented as areas, which are generally known as polygons.

Features generally have information associated with them, which in GIS nomenclature we term attributes. In our example, all the features have an attribute that specifies their name, as well as geometric information such as road length or county area. These attributes are stored in a table that is linked to the feature, meaning we can easily access this information for any feature, as well as performing search queries. For example, we might want to find all the motorways in our roads layer or all

Spatial Data Models

counties over a specific population size. GIS permits us to do this. We can then use this information to create new data layers where only our specific criteria are met, such as a separate layer of motorways or counties with a high population density. Similarly, we could color all the road types a different color to allow us to distinguish them. We learn how to create new data from these attribute queries in Chapter 5 and how to change the symbology to create simple location maps in Chapter 7.

Another example of attribute data would be statistics related to the feature. For example, if our features were all the counties of Ireland, we could have the following type of attribute data, county area, county population, or number of people unemployed. We could use this attribute data to shade or color the counties and produce a thematic map. Using the population statistics, we could create a thematic map showing counties with larger populations in dark red and those with smaller population in light red; this type of map would allow us to see the heavily populated areas of Ireland very quickly. We develop thematic maps in Section 3.

10. Now that we have imported data, it is time to save our project file. We do not necessarily want to repeat the previous steps should something go wrong. Save the overall project as Chapter 3.

Now it is time to check out the attribute information

11. Right click on the layer Counties, and select Open Attribute Table.

A table that should resemble Figure 3.6 appears. This is the 2011 census data at county level for Ireland with information on the 2011 total population.

Another hint: When working within a GIS, the attribute table is an integral part of any workflow. Therefore, we often want it in permanent view. If we look on the toolbar within the attribute table, we should see the tool to dock this. If we click it, we

	NUTS1	NUTS1NAME	NUTS2	NUTS2NAME	NUTS3	NUTS3NAME	COUNTY	COUNTYNAME	Total2011	Shape_Leng	Shape_Area
1	IE0	Ireland	IE01	Border,Midland...	IE011	Border	33	Donegal County	161137.0000000...	1532962.130230...	4862655102.670...
2	IE0	Ireland	IE01	Border,Midland...	IE011	Border	34	Monaghan Cou...	60483.0000000...	293372.6410499...	1294706775.019...
3	IE0	Ireland	IE02	Southern and E...	IE024	South-East (IE)	01	Carlow County	54612.0000000...	220379.2810820...	896637626.9440...
4	IE0	Ireland	IE02	Southern and E...	IE021	Dublin	02	Dublin City	527612.0000000...	96306.61053980...	117702315.4689...
5	IE0	Ireland	IE02	Southern and E...	IE021	Dublin	03	South Dublin	265205.0000000...	91784.17730500...	223507436.4990...
6	IE0	Ireland	IE02	Southern and E...	IE021	Dublin	04	Fingal	273991.0000000...	232028.7397910...	458211861.4200...
7	IE0	Ireland	IE02	Southern and E...	IE021	Dublin	05	DÃ/Æ Ã Ã°n La...	206261.0000000...	67003.58808410...	126630761.7859...
8	IE0	Ireland	IE02	Southern and E...	IE022	Mid-East	06	Kildare County	210312.0000000...	292336.0868829...	1694921385.380...
9	IE0	Ireland	IE02	Southern and E...	IE024	South-East (IE)	07	Kilkenny County	95419.0000000...	281223.0720749...	2072030598.730...
10	IE0	Ireland	IE01	Border,Midland...	IE012	Midland	08	Laois County	80559.0000000...	271669.2237740...	1719862555.950...
11	IE0	Ireland	IE01	Border,Midland...	IE012	Midland	09	Longford County	39000.0000000...	211099.8058620...	1091620196.079...
12	IE0	Ireland	IE01	Border,Midland...	IE011	Border	10	Louth County	122897.0000000...	247192.2863830...	827121687.7150...
13	IE0	Ireland	IE02	Southern and E...	IE022	Mid-East	11	Meath County	184135.0000000...	386616.2330259...	2342991251.960...
14	IE0	Ireland	IE01	Border,Midland...	IE012	Midland	12	Offaly County	76687.0000000...	383748.9032610...	2000890039.960...
15	IE0	Ireland	IE01	Border,Midland...	IE012	Midland	13	Westmeath Co...	86164.0000000...	287040.7147199...	1839119058.470...

FIGURE 3.6 Overview of the attribute table for the CSO census data affiliated with the counties layer.

create a tabbed interface where the attribute table can be viewed without obstructing the map view, but more importantly it can be viewed on a permanent basis. I find this quite useful, as if the attribute table is not docked, when you return to the main map canvas the attribute table remains open but hidden as another QGIS window. However, if you do not like this view, no problem! Each GIS journey will be unique. To undock the attribute table, simply click on the same button again.

Now we have opened the attribute table, we can use it to explore the values associated with the spatial data. One question we might be interested in answering is to find the county that has the highest total population.

12. Click on the column heading Total2011.

It has ordered it in ascending order. We can either scroll to the bottom of the attribute table or click the column heading again to order it in descending order. The county with the highest population in 2011 was Dublin City with 527,612.

13. Click on the gray box to select the first record in the attribute table, it will turn blue to indicate it is selected.

Dublin City is also highlighted in yellow. The attribute table and features are linked, so that when we select a row in the attribute table, the feature is also selected. If you cannot see this, turn off the Roads and Schools layer.

14. Explore the attribute table, selecting different counties. Hint: we can select more than one county if we hold Shift or Ctrl.
15. Turn back on the layer of Schools if you have turned it off.
16. Click on the Zoom-In button and draw a square around the Dublin area, as shown in Figure 3.7.

This should have zoomed us into Dublin. We are not actually drawing a permanent box, but simply specifying the extent of the zoom. Do not worry if you have not achieved this perfectly the first time, we are going to try again.

17. Click on the Zoom-Full button. This button automatically displays the map at the extent of the largest layer that is in the GIS.
18. Try zooming in on Dublin again.

Hopefully this should be zoomed in perfectly on Dublin. If not, try again.

19. Instead of the Zoom-Full button, try the Zoom-Out button.

Consider these buttons. I often find the Zoom-Out button difficult to control and find that it zooms us out at a scale that is defined by a rectangle that is not always intuitive. I try to use a combination of the full extent and zoom-in, but over large geographic areas this is not always feasible. You will have your own preference for tools, so use the tools that feel the most comfortable. It can also be a lot simpler to use the

Spatial Data Models

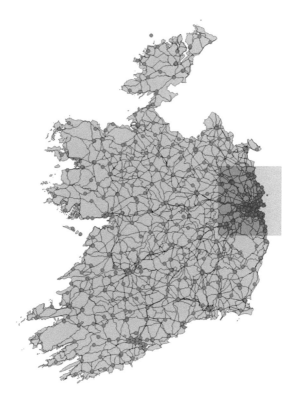

FIGURE 3.7 Screenshot of the zoom tool in use around the Dublin area.

scroll pad on the mouse, so play around and use the function that suits you the best. Regardless of the method, we should notice the map scale changing in the bottom panel of our QGIS interface. This is the ratio of a distance in the real world compared to the same distance on the map.

20. Now, let us use the Pan Map tool to pan around the country without changing the zoom.
21. Click on the Identify Features button, and select a random school.
22. A dialogue box opens with information on the official school name, as well as some other information.
23. When you have adequately explored the data and are ready to move on, click on the Zoom-Full to return to the full extent.

We are now going to add some climate data into the GIS layering. There are two current projections of climate and one set of future projections that have been obtained from Hijmans et al. (2005). There are 19 bioclimatic variables in total, each representing different climatic conditions. The two that we work with in this chapter are:

- Bio6 – minimum temperature of coldest month
- Bio13 – precipitation of the wettest month

FIGURE 3.8 Screenshot of the Climate_bio6 raster layer.

The future projection represents the Shared Socio-Economic Pathway 585, which is the worst-case future projections regarding climate change.

24. Navigate through the tab Layer > Add Layer > Add Raster Layer.
25. Add the Climate_bio6.tif layer.

Your map should resemble Figure 3.8. Attributes are stored in raster datasets slightly different, meaning there is no attribute table.

26. Click on the Climate_bio6 layer in the Layers Panel so that it is selected.
27. Use the Identify Features button to click anywhere on the raster.

This should open a small dialogue box that is either floating or docked under the processing toolbox. The Identify Results Panel should have opened. This tells us that the raster contains one band, or variable. This represents the minimum temperature in °C of the coldest month from 1960 to 1990.

28. Next, add the Climate_bio13 layer, using the same steps as just implemented for Climate_bio6.

Spatial Data Models

FIGURE 3.9 Screenshot of the Climate_Future raster layer.

It is quite hard to discern any pattern between the two layers as we must turn on and off the different layers in the Layers Panel to view them. Despite that, we do observe some patterns, particularly the colder temperatures in the north and east and the wetter areas in the south and west. Next, let us observe how these layers are projected to change.

29. Add the layers representing future climate scenarios, Climate_Future.

The first thing we notice is that the raster layer looks quite different, in fact quite psychedelic, as shown in Figure 3.9. We may have also noticed that we have only added one layer. This is because we can represent the real-world using rasters that have a single band (such as those representing the current climate) or rasters that have multiple bands. In our case for future projections, we have a raster that has 19 bands (that represent each of the 19 bioclimatic values). When we have multiple bands, each location has more than one value associated with it, and is useful for a lot of datasets, including remotely sensed imagery that relies on the electromagnetic spectrum. It also provides an efficient way of storing multiple datasets, as it has in our case. We can choose to display multiple layers combined or display a single band. For our example, it does not make sense to combine the bands, meaning we want to visualize just the two bands we are interested in.

30. Right click on the Climate_Future layer, and select Properties.
31. Click on the Symbology tab.
32. Change Render Type to Singleband Gray.
33. We are interested in exploring the coldest temperatures, so choose Band 06: wc2_6 under Gray Band, and then click Apply.

This should now change the color band to a single black to white scale. We should investigate how our representation of future minimum temperature compares to the current minimum temperature (Climate_bio6). Hint: we may need to turn off some layers, and/or rearrange the drawing order in the Layers Panel.

Despite that, it is quite hard to discern any substantial changes. Here we can use the overlay functionality that can be used to quantitatively assess the amount of change in the minimum temperature. To do this, we use the Raster Calculator.

34. Navigate through the tab Raster > Raster Calculator.

Here we are specifying an expression that will allow us to quantitatively overlay the two layers. Firstly, we have to specify the layers, then the operation between them.

35. Scroll to Climate_Future@6 and double click on it. It should appear in the expression box.

Remember, this layer is a composite of 19 bands, so we are selecting band 6 which is the minimum temperature of the coldest quarter. The @ sign specifies that this is the relevant band. When we do this for the current climate, raster calculator still requires us to specify the band, even if there is only one.

36. Click on the minus sign (–).
37. Then scroll to Climate_bio6 and double click on it. The expression should look like the following:
    "Climate_Future@6" – "Climate_bio6@1"
38. Save the output layer as a permanent layer by clicking on ... by the Output layer and saving in your working directory. If your dialogue box is grayed out, most likely it will be an issue with where you have specified the save. I will call this Bio6_Difference.
39. Ensure your dialogue box resembles Figure 3.10. Click OK.

Our output should resemble Figure 3.11. The values range from 0.6 to 4.1, meaning in all locations in Ireland, the minimum temperature is going to get warmer by at least 0.6°C. We can also see that it is going to get particularly warmer along the northern coasts (as lighter colors indicate higher values) and in pockets within the midlands. The maximum deviation of temperature will be in the south-east region where it will get warmer by 4.1°C.

Returning to the SDG aim of education and climate mitigation, we can now track the projected climatic changes at each individual school. We could pan/zoom to a school and investigate what the temperature change will be. If we consider ourselves

Spatial Data Models

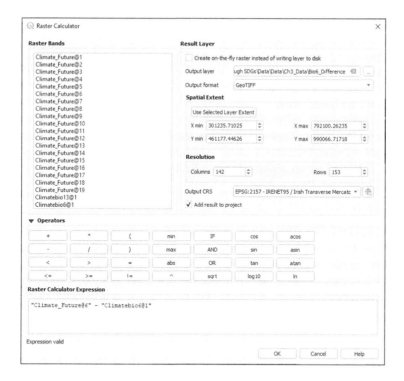

FIGURE 3.10 Screenshot of raster calculator to generate difference in Bioclim 6.

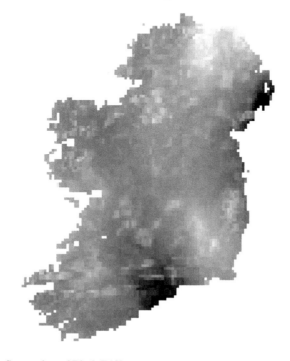

FIGURE 3.11 Screenshot of Bio6_Difference raster.

part of the education team there, considering how to integrate climate mitigation into the curricula, one option would be to compare our current climate with that projected in the future and provide the baseline data. Currently, we could do this qualitatively by exploring the raster values that underlie each school, but we can extract these values to the point features that represent schools. To do this, we are going to use the Sample Raster Values tool.

40. If the Processing Toolbox is not open, either click on the Processing Toolbox button in the Attributes Toolbar or turn it on by right clicking the gray area. The tool we are looking for is in Raster Analysis.
41. Navigate to Raster Analysis > Sample Raster Values. Alternatively, we could type the tool name in the search panel.
42. Set the Input Layer as Schools.
43. Set the Raster Layer as Bio6_difference.
44. Specify the output column prefix as 'bio6'. This is what the new attribute we are creating will be called.
45. In Sampled, click on the … to save the new layer as a file in your working directory as 'Schools_bio6Diff'. For now, save it as a shapefile (we revisit data formats in Chapter 6).
46. Ensure your dialogue box resembles Figure 3.12. Click Run.

We now have a new layer of schools. Let us check whether the values have been appended.

47. Open the attribute table.

FIGURE 3.12 Screenshot of the parameters needed for the Sample Raster Values tool.

Spatial Data Models

There is one new column in the attribute table (bio6), representing the temperature difference. We can sort this column to identify the school that has the largest projected increase in temperature. For reference, this is on the north coast.

48. Use these new GIS skills to repeat steps 34–47 for precipitation. Remember, precipitation is the Bioclim 13 layer. When undertaking step 37, ensure that your raster calculator statement includes 13 and not 6 to reflect precipitation and not temperature. Similarly, when undertaking step 42, choose the newly created Schools_bi6diff layer as the Input Layer, this way both values append to the point layer.

Again, for reference the school with the highest projected increase in precipitation of 115 mm is in the northern most county, Donegal, and there are four schools in the south-west of Ireland (near Tralee) that have a projected increase of 81 mm. If you have negative values, check that the rasters have been input in the correct order in Step 37. Given the target of SDG13 is to educate, here we focused on schools as the feature of interest; however, there is no reason why this cannot be any other real-world entity. Consider, for example, we are working for the government, and we have been tasked with identifying roads at risk of increased surface water under future precipitation events. We could use the steps within this chapter and calculate the largest increase in precipitation for the country and sample those values for the roads layer.

However, the Sample Raster Values tool relies on our features being represented as points. The roads layer is represented as lines, meaning we need to find another tool to complete this. This is not unusual in GIS, as many of the conceptualizations and methodologies we implement are broadly similar across spatial data models; however, the specific tools that have been developed work only for a specific type of feature. Therefore, we can use two other tools to extract the raster information to our roads layer. The two tools are Drape and Extract Z values.

Drape extracts the raster value to each vertice in the road layer. We know that in GIS, all lines are captured by points, connected precisely by straight lines. All points within a line file are known in GIS terminology as vertices. I should note here that these vertices are not visible in the GIS graphic user interface, but they are present. Therefore, this tool extracts the raster values to each of these vertices. The Extract Z values use this information to create a new attribute in the feature layer, such as the average of all values within a feature's vertices. Therefore, we can use these two tools to capture the information contained within the precipitation raster.

49. In the processing toolbox, navigate to Vector Geometry > Drape (set Z value from Raster).
50. Set Roads as the Input Layer.
51. Set Bio13_Diff as the Raster Layer.
52. Save the output Draped file as Drape_PPT (again, we can save this as a shapefile).
53. Keep all other options as default.
54. Ensure your dialogue box resembles Figure 3.13. Click Run.

FIGURE 3.13 Screenshot of the parameters needed for the Drape tool.

This should have created a new road layer. Next, we want to summarize all the information that has been generated for each vertice.

55. Navigate to Extract Z values in the processing toolbox.
56. Set the Input layer as Roads_PPT.
57. Click on the ... for summaries to calculate.
58. Ensure that Mean is the only option ticked. This will generate the mean precipitation difference from the raster layer for the vertices of that line.
59. Save the column prefix as PPT.
60. Save the output layer in the Extracted box as Roads_PPT in your working directory (again as a shapefile).
61. Ensure your dialogue box resembles Figure 3.14. Click Run.

We can now open the attribute table and explore the road feature that will be subject to the largest increase in precipitation during the wettest month. For reference, this is also in Donegal in the north of Ireland.

62. In the Browser Panel, navigate to XYZ Tiles and turn on OpenStreetMap to provide reference.

This information will support governments to make informed decisions regarding which roads are most likely to be impacted by climate change, specifically increased precipitation events that may require targeted infrastructure modifications such as culverts or long-term water storage facilities to reduce flash flooding in times of high rainfall. For this specific road, there may need to be more targeted adaptation strategies related to the river Finn that runs parallel to the road.

Spatial Data Models

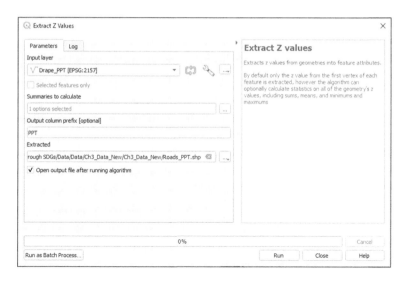

FIGURE 3.14 Screenshot of the parameters needed for the Extract Z values tool.

Finally, I mentioned at the start of the chapter that discrete objects (i.e., schools) are more often represented using the vector data model and continuous fields (i.e., temperature) using the raster data model; however, such codifications are not exclusive. The final stage of this chapter will demonstrate some of the conversion tools that are available within GIS, should we wish to convert from vector to raster or raster to vector.

63. Navigate through the tab Raster > Conversion.

There is a list of tools that can be used to transform data across spatial data models. Firstly, we will convert our school data to raster format.

64. Select the Rasterize (Vector to Raster) tool.
65. Select Schools as the Input Layer.
66. Select the fixed value to burn, which will give the continuous raster a value where we have a school. As this layer will represent a categorical consideration of schools, we can use 1.
67. Set the output raster size units to 'georeferenced units'.
68. Set the width and height of the resolution to 1000.

As raster layers are a continuous gridded representation, we need to define the vertical and horizontal height of each unit so that it is consistent across space. The last two commands have specified that the output raster will be calculated using the units of the map document, with a distance of 1000 units. The data is projected using a Cartesian coordinate system using meters as the unit measurement. Therefore, we have created a 1000 m gridded representation of schools or no schools. We revisit projections in the next chapter.

69. Select the Output Extent options ... and select from Layer, choosing the same extent as Climate_bio6.
70. Save the Output layer (Rasterized) in your working directory, I'm calling mine School_Raster.
71. Ensure your dialogue box resembles Figure 3.15. Click Run.

We now have a gridded representation of schools in Ireland. If we turn all layers off except for the two school layers, they should align. Next, we are going to perform the opposite for Climate_bio6, and vectorize the data.

72. Navigate through the tab Raster > Conversion > Polygonize (Raster to Vector).
73. Select Climate_bio6 as the Input layer.
74. Change the name of field to create to Temperature. DN is the default option, which refers to digital number.
75. Save the Vectorized layer as a shapefile in your working directory, I'm calling mine Temperature_Polygon.
76. Ensure your dialogue box resembles Figure 3.16. Click Run.

FIGURE 3.15   Screenshot of the parameters needed for the rasterize tool.

Spatial Data Models 39

FIGURE 3.16 Screenshot of the parameters needed for the polygonize tool.

Note, you may receive an error stating that your field name 'Temperature' has been changed to 'Temperatur'. This is because shapefiles can only save field names of up to ten characters. More on this in Chapter 6. Because of the method of conversion, all the decimal values have been lost from the raster dataset, meaning our new polygon only ranges from −1 to 6. We could have overcome this prior to conversion by using the raster calculator and multiplying temperature by 100 to ensure that we had full integers before vectorizing. However, the main point of this final exercise is to illustrate that the same data can be represented using both spatial data models and not to fall into the trap that can easily happen when starting out in GIS that specific features must be represented using a certain spatial data model. As we work through the book, we encounter several instances where we must convert the data from one format to another, with many of these decisions falling squarely on our shoulders as the GIS operator. Therefore, understanding just what is possible is important, and this is where we must think critically about how we can use spatial data models to represent our geographic features. We can clearly see the differences between the vector and raster models for schools and temperature, as shown in Figure 3.17, but acknowledge that it is possible to represent both features using both models. Please note that I have changed the color palette in Figure 3.17 for presentation purposes, so your outputs may look slightly different.

3.3 CASE STUDY CONCLUSIONS

Through layering our data in GIS and representing the real world through different spatial data models, we have been able to combine discrete objects (i.e., schools, roads) with continuous fields (i.e., temperature and precipitation). We

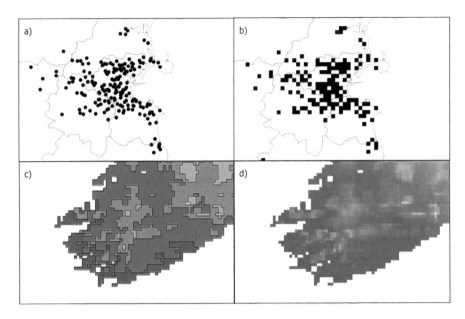

FIGURE 3.17 School data presented in (a) vector and (b) raster formats. Temperature data presented in (c) vector and (d) raster formats.

have been able to represent the difference in continuous variables as a raster layer, but also extract this information to point format, highlighting some very basic analysis that GIS can implement in supporting climate mitigation and resilience education.

We have now reached the end of the first case study chapter. We have explored the two predominant spatial data models used in GIS, vector and raster, and used the information stored within these layers to support SDG13, specifically investigating the impact of climate change on education institutes and government infrastructure. In doing so, we have layered the GIS data to visualize this information, as well as link the spatial data with the attribute data. We have used several of the toolbar buttons for navigating through the QGIS interface, and implemented spatial operations, creating new data that incorporates information from multiple layers. Subsequently, we were able to identify which schools and roads are projected to have the largest changes under future climate scenarios, which could be used to mitigate the impacts of climate change. Finally, we converted data from raster to vector and vice versa. These final steps were not used in our case study, but such transformations were important to highlight early in our GIS journey and we revisit these conversion tools in the second half of this book. It is important to remember that spatial data models are abstractions of reality, and as such we should always think critically about which ones to use, while being aware of the tools to convert should we deem it appropriate. As we progress through this book, we continue to reinforce the learning outcomes completed in this chapter and build upon these skills.

3.3.1 Test Yourself

If you want to test yourself on the learning outcomes of this chapter, complete the following:

a. Identify the schools and roads that are projected to have the smallest changes in temperate and precipitation, respectively.
b. Convert the vector dataset of roads to a 100 m raster format.

REFERENCES

All-Island Research Observatory (AIRO) (2016). *Post Primary Schools [dataset]*. Available from: https://data.gov.ie/dataset/post-primary-schools?package_type=dataset. Accessed March 1, 2022.

Central Statistics Office (CSO) (2013). *Census 2011 Boundary Files*. Available from: https://www.cso.ie/en/census/census2011boundaryfiles/. Accessed March 1, 2022.

Hahmann, S. and Burghardt, D., 2013. How much information is geospatially referenced? Networks and cognition. *International Journal of Geographical Information Science*, 27(6), pp. 1171–1189.

Hijmans, R.J., Cameron, S.E., Parra, J.L., Jones, P.G. and Jarvis, A., 2005. Very high resolution interpolated climate surfaces for global land areas. *International Journal of Climatology: A Journal of the Royal Meteorological Society*, 25(15), pp. 1965–1978.

Ordnance Survey Ireland (OSi) (2016). *Roads – OSi National 250k Map of Ireland*. Available from: https://data-osi.opendata.arcgis.com/datasets/1434c3b05da742cdb47e00040edc9dd5_24/explore?location=53.281779%2C-8.238600%2C7.72. Accessed March 1, 2022.

4 Projections

4.1 INTRODUCTION AND LEARNING OUTCOMES

As outlined in the previous chapter, the basic elements of spatial data include the location of features (i.e., the coordinates), the attribute data (i.e., the phenomena of interest), and a transformation to convert geographic locations into a position on a flat surface (i.e., a projection). In this chapter, we cover projections, detailing just how vitally important this element of spatial data is. We investigate how each of these projections distorts the world map and how to choose the appropriate projection for different tasks.

The simplest analogy to describe what we are trying to achieve with coordinate systems is to imagine we are peeling an orange. In one go. Without breaking the peel. If we manage to achieve this (in some cases superhuman) feat, we cannot simply place the unbroken peel on a flat surface and have a perfectly square representation of the orange's surface. It is a rugged, unique, and disjointed orange peel. In our case, the Earth is our orange, and we are trying to represent the three-dimensional curved surface of Earth on a two-dimensional flat surface. To create such continuous representations of space (either through maps or databases), we need to stitch these locations together. Subsequently, transforming our spatial data into two dimensions is by no means a simple task, although many modern-day GIS software can certainly make it seem simpler than perhaps it is (however, we get to this in due course). In other words, we are trying to represent the Earth first as a three-dimensional sphere, before transforming it further to a two-dimensional flat surface. To achieve this, first we need to define the Geographic Coordinate System (GCS) and then the Projected Coordinate System (PCS), with a schematic overview provided in Figure 4.1.

The GCS is based on the premise of a perfect sphere and uses datums to define the mathematical properties of the shape. If we consider a perfect sphere, we can identify imaginary circles covering the entire shape. All circles whose circumference runs in a north to south direction are termed 'great circles' or 'meridians'. These circles are all the same size, and represent the largest circumference of the sphere, hence the term 'great circles'. All circles whose circumference follows an east to west direction are termed 'small circles' or 'parallels'. These circles are widest at the center of the sphere (the equator), and then get progressively smaller as they approach the top and bottom of the sphere (the poles). These circles, termed graticules, are of course not actually present on the globe but represent what we term longitude and latitude and the method by which we position ourselves on the three-dimensional sphere.

Longitude is used to locate our east-west position on the Earth's surface and consists of the meridians. As such, there is no natural origin. In 1884, the Observatory of Greenwich in Great Britain was adopted as the official prime meridian (or 0° longitude), with degrees of longitude measured from 0° to 180° east and 0° to 180° west from this location. The location of the prime meridian has been the focus of several

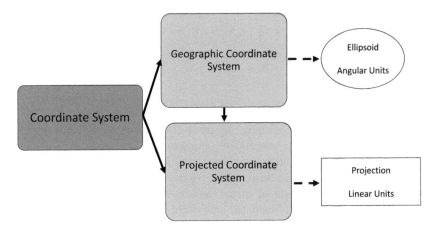

FIGURE 4.1 Overview of coordinate systems for use in geographic information systems (GIS).

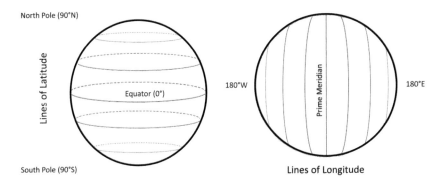

FIGURE 4.2 Diagram of latitude and longitude.

discussions regarding geographies of power, stemming from the colonial history of Great Britain, meaning in theory, any of the 360 meridians could be 0°. Such discussions are central when considering the role maps have played in the history of society, especially the geopolitics of centering countries in the middle of maps; however, such a concept is equally important when we come to project our three-dimensional sphere onto a flat surface, and we revisit that later in the chapter. Latitude is perhaps simpler, with the largest circle at the equator, representing 0°, and each subsequent angular degree increasing until the poles are reached for 90° north and 90° south. Figure 4.2 provides an overview of latitude and longitude.

Latitude and longitude are circles and so best represented using angular units. They are traditionally measured in degrees, minutes, and seconds (DMS). Degrees are expressed as angular units, and simply work by taking the angle between the equator and the location on the sphere's surface (although do note there are deviations to this method). Minutes and seconds are used to store digital coordinate information and subsequently expressed as decimal values. If we take the below example, we can see that the location is 34° 33′ 30″, meaning we know that the location falls between the angles 34 and 35. To calculate the value of minutes and seconds, we need to divide the value by the number of minutes and seconds in a degree, which we can think of as analogous to an hour. There are 60 minutes and 3600 seconds in an hour, meaning:

- 33/60 = 0.55°
- 30/3600 = 0.00833°

Before we simply sum them together:

- 34 + 0.55 + 0.00833 = 34.55833DD

Therefore, the concept of identifying the location on the Earth's surface in three dimensions is predicated on the idea of a perfect sphere; however, the Earth is only roughly spherical. For our purposes, be that mapping, surveying, engineering, or whatever, we need to be much more precise. There are visible variations such as mountains and valleys, but deviations from a perfect sphere would still exist if all the mountains and valleys were flattened out. We can build on our fruit analogy here to support our understanding. First, if we start with our sphere, or as alluded to earlier, our orange, we slightly flatten it until we have a shape resembling a grapefruit. However, the shape is slightly narrower in the northern latitudes, while it expands in the southern latitudes, which would somewhat resemble a squashed pear. This shape is what we refer to as an ellipsoid. Finally, if we add some local bulges in sea level to the shape, or what would be sea level if the whole planet was ocean, then we are left with a potato shape, which is termed the geoid. I should note here that the analogies to the distortions have been greatly exaggerated for understanding. The Earth is not really a floating giant potato, but the comparisons can support a level of understanding in the differences.

The level of flattening in the ellipsoid between the polar radius and equatorial radius is about 21.5 km, making the circumference of the ellipsoid approximately 40 million meters. The Earth rotates around its shortest axis, which is the north-south axis, making the Earth an oblate ellipsoid. The ellipsoid model is necessary for accurate range and bearing calculations over a long distance, such as Global Positioning Satellite (GPS) navigation. Although the Earth's shape is technically an ellipsoid, its major and minor axes do not vary greatly. Because of this, the term spheroid and ellipsoid are often used interchangeably in GIS literature and software, but there are important differences between a perfect sphere and an ellipsoid. This is because different countries use different rates of flattening when calculating the degree of oblateness in the ellipsoid as local variations exist, which means different ellipsoids perform differently at locally approximating the geoid. This is where the datum comes into play.

The datum defines the shape of the ellipsoid and the origin and orientation of the ellipsoid to the geoid and is often referred to as a reference mapping system. There are datums for different parts of the world, based on different measurements, meaning there is a large diversity of datums. Assigning the wrong datum to a coordinate

FIGURE 4.3 Conceptual diagram of the different ellipsoids that can be used to best fit different locations on the globe. In the left-hand example, the ellipsoid best fits North America and South America, in the right-hand image the ellipsoid best fits Europe and Africa, while in the center image, the ellipsoid is the best fit for the whole Earth.

system may result in errors of hundreds of meters, which can have substantial implications on the outputs we work on as GIS specialists. The origin is the point where the ellipsoid matches up perfectly with the surface of the Earth, and where the latitude and longitude coordinates on the ellipsoid are true and accurate. All other points on the surface are referenced to this location. In other words, a datum describes how the geographic coordinate system assigns latitude and longitude values to feature locations.

The geoid is subsequently the true three-dimensional shape of the Earth, where it is considered as mean sea level extended continuously through continents, which is the result of a surface of constant gravitational potential. For example, in Figure 4.3, it can be seen how different ellipsoids are preferable in different locations due to their ability to preserve accurate measurements in different locations. The datum subsequently aligns the ellipsoid to the surface of the Earth in different regions.

To recap, the datum defines:

- The ellipsoid
- The orientation of the ellipsoid to the geoid
- An origin
- A direction from the origin to another location

For most applied GIS work, the process of choosing and selecting the ellipsoid and datum is often well established and a decision that has been long made. Such considerations are much more central in the data collection phase of research, particularly in disciplines such as engineering; however, that does not mean that (1) we do not need to know which datum has been used, and (2) how such considerations impact our analysis. Later in this chapter, we explore how to identify such information when provided with data. There are subsequently various possible locations when provided with coordinates in latitude and longitude, and the datum/ellipsoid that is used will impact the location of our features. At the very least, something to think about next time we are watching a movie about hidden treasure buried at historical latitude and longitude measurements...

A map projection (or hereafter simply projection) is the method of representing data located on a three-dimensional surface onto a flat plane. We can use projections with different datums; however, every projection causes distortion as it tries to convert

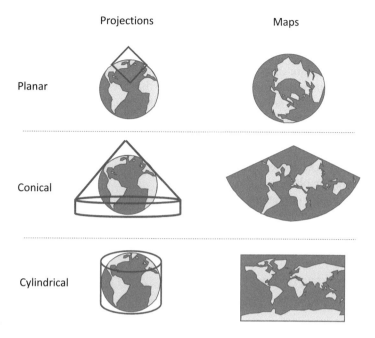

FIGURE 4.4 Diagram of the three predominant projection types: planar, conical, and cylindrical, and their corresponding map surfaces.

spherical information onto a flat surface. Computationally, it is much easier to work with Cartesian coordinates, which are defined in linear units (i.e., meters and feet), and referred to as easting and northings; however, a projection will distort at least one of the following elements: shape, area, distance, or direction. Therefore, it is our job as GIS users to understand these distortions and determine which parameter is important.

There are three main types of projections: cylindrical, conical, and planar. These are presented in Figure 4.4. In cylindrical and conical projections, the reference spherical surface is projected onto a cylinder wrapped around the globe, and then cut and unwrapped to form a flat map. Cylindrical projections tend to align with the meridians, while conical projections tend to align with the parallels. Where the projection 'touches' the reference surface (often along one of these circles) there is no distortion, and the scale factor is 1. These are referred to as lines of tangent or secant. Tangent lines are where there is only one location where the projection touches the reference globe, while secant lines are where the projection cuts the surface, meaning there are two locations. In cylindrical projections, these are lines of equidistance, and distortion increases by moving away from these standard lines. In conical projections, features often appear larger as distortion increases. In planar (or azimuthal) projections, the reference surface is projected onto a plane, where direction is always true from the center.

There are four main map types:

- Conformal, which preserves local shape
- Equivalent, which preserves local area

- Equidistance, which preserves length
- Azimuthal, which preserves direction

Maps can have more than one property, but conformal and equivalent are mutually exclusive, meaning we cannot have a map that preserves both area and shape. We will now explore some of these properties within QGIS. By the end of this chapter, you will have completed three learning outcomes, and you should be able to:

- Explain how choosing a projection can impact our understanding of the geographic phenomena we study.
- Perform GIS operations related to defining and re-projecting the coordinate systems for spatial data.
- Perform geometric operations to measure length and area using different coordinate systems.

4.2 CASE STUDY: MEASURING FIRE SIZE TO SUPPORT THE SDGs

In this chapter, instead of focusing on one SDG, we investigate a phenomenon that is ubiquitous to them all: Fire. While fire is not explicitly mentioned within the SDGs, Martin (2019) argues for its inclusion on the basis that the effects of fires can have a far-reaching impact on many social, economic, atmospheric, terrestrial, and marine services. For example, SDG2.4 aims to ensure sustainable food production systems and strengthen the capacity for adaptation in response to disasters. Similarly, SDG3 aims to ensure healthy lives and wellbeing, with fire often associated with a reduction in air quality, a conversion in land cover, and potential damages to cultural artifacts. SDG15 aims to protect, restore, and promote sustainable use of terrestrial ecosystems, with fires causing the combustion of vegetation and conversion to other types. Savannas are mixed-plant communities, comprising of grass and woody vegetation, covering approximately a quarter of the Earth's surface, and over half of the African continent. They are an extremely important socio-economic landscape, supporting a diversity of societies, communities, and biodiversity. Fire has a nuanced relationship with vegetation dynamics in savanna ecosystems (Marden et al. 2022), in some cases preventing closed canopy forests from forming, while in others leading to bush thickening of often-inhospitable grazing plants, degrading the quality of life for pastoralist societies (Meyer et al. 2019). Therefore, simply knowing the burnt area of protected locations could provide targeted support to any number of stakeholders across the SDGs.

Woods and Govender (2004) released historical fire maps of Kruger National Park in South Africa over a ten-year period from 1992 to 2001. This dataset was published and openly shared, without restriction in accordance with EOSDIS Data Use Policy. Here we use this dataset to investigate how projections can distort the data we are working with, as well as completing some basic GIS skills such as measurements. All other data used in this chapter has been generated by the author.

Projections

1. Open QGIS and start a New Empty Project. If you are working on the same device, you should now see Chapter 3 in your recent projects.
2. Navigate through the tab Layer > Add Layer > Add Vector.
3. Add the knp_fires_1992–2001 shapefile.
4. Next, in the Browser Panel, navigate to XYZ Tiles and double click on OpenStreetMap.
5. Reorder the layers such that the OpenStreetMap is underneath the knp_fires_1992–2001 layer in the Layers Panel.

Your QGIS interface should look like that in Figure 4.5.

A polygon dataset that appears to run the length of the eastern border of South Africa should be visible. The first thing that should strike us upon investigating the pattern is that it appears that during the ten-year period between 1992 and 2001, almost all of Kruger National Park experienced some form of fire.

6. Before we begin, let us save our project should something happen. Save the project as Chapter 4. Remember to save intermittently throughout the chapter.

One of the first things we should always do when working with GIS data is to check the datum and projection. This is important, as for us to implement any analysis, we need to ensure that all data is projected in the same coordinate reference system (CRS).

7. Right click on the knp_fires_1992–2001 layer in the Layers Panel and select properties.

FIGURE 4.5 Screenshot of fire layer in Kruger National Park, South Africa, 1992–2001. Basemap is the OpenStreetMap XYZ tiles which is © OpenStreetMap contributors and available under the Open Database License. Please see https://www.openstreetmap.org/copyright.

This should open a dialogue box with various tabs.

8. Click on the Information Tab.

Your dialogue box should look like that presented in Figure 4.6.

The CRS is clearly outlined in the information tab, and we can observe that the data layer is projected using a WGS84 datum and projected from the ellipsoid using UTM zone 36S.

Universal Transverse Mercator (UTM) is a very common coordinate system that is based on the Transverse Mercator projection. This is a projection that uses a cylinder turned on its side. UTM divides the Earth into 60 bins of 6° of longitude. These are then further delineated into the northern and southern hemispheres between 84°N and 80°S, respectively. For each 6° zone, a Transverse Mercator projection is applied to the zone's central meridian. For example, Zone 20 is between 66°W and 60°W, meaning the central meridian for this zone is 63°W. This is based on the aforementioned fact that each of the meridians are the same size with no natural origin. In each UTM zone, the central meridian has two secant lines either side of it to minimize distortion, meaning the central meridian has a scale factor of 0.9996. This approach has been adopted to ensure maximum accuracy across each zone, and

FIGURE 4.6 Screenshot of the information tab in properties to identify the coordinate reference system of the knp_fires layer.

subsequently the globe. This process is repeated for each of the 60 zones, resulting in a stitched together coordinate system that minimizes error at a global scale, and we can locate ourselves anywhere on the Earth using UTM.

To ensure that no coordinate has a negative value, we first separate the UTM zones into northern and southern hemispheres and then apply a false origin to the west of the central meridian of 500,000 m. This means the value of the central meridian in the zone will always be 500,000 m, and any values east or west of this will always be positive. Because the circumference of the ellipsoid is approximately 40,000,000 m, we can assume that each zone has 20,000,000 m in the N-S direction. For the northern hemisphere, northings start at the equator and extend to 10,000,000 m. For the southern hemisphere, to ensure no negative values, the equator has a value of 10,000,000 m, with values subtracted from it. Therefore, to locate ourselves using UTM, firstly we identify the zone we are in, then we calculate the easting and then finally the northing.

Throughout the chapters, we mainly project data using UTM, as our case studies are global in their reach. For example, in this chapter the UTM zone is 36S, meaning we are in zone 36 in the southern hemisphere. This is important to note, and any data we subsequently add needs to be in this projection for our data to overlay and for us to be able to undertake efficient analysis. This is the primary reason we always check the projection of our data when first opening it in a GIS.

The second thing we should do when working with new data is to open the attribute table so we can fully understand the data we are using.

9. Right click on the layer in the Layers Panel and select Open Attribute Table (you may wish to dock the table as we explored in the previous chapter).

There are several attributes that can be used to enhance the spatial data, including the date and time of the fire, what the cause was, who the agent was, as well as the intensity of the fire. If we scroll across to the end of the attribute table, the final two columns detail the area and the perimeter of the fire burn.

10. Click on the attribute 'AREA_HA'.

This will sort the area in hectares from smallest to largest. If we click again, it will sort from largest to smallest. The largest fire has an area of 204,312 ha. This is shown in Figure 4.7.

11. Select the row in the attribute table for the largest fire during the period. This links the data to the map and identifies where this fire occurred.

This burned area appears in the south of the national park but appears to be made up of multiple polygons. This is not actually the case, but the result of several polygons, or fires, that are overlapped or stacked during the decade. To identify the overall location of the largest burned area, we are going to export it to its own layer.

12. Right click on the knp_fires_1992–2001 layer in the Layers Panel and select Export > Save Selected Features As.

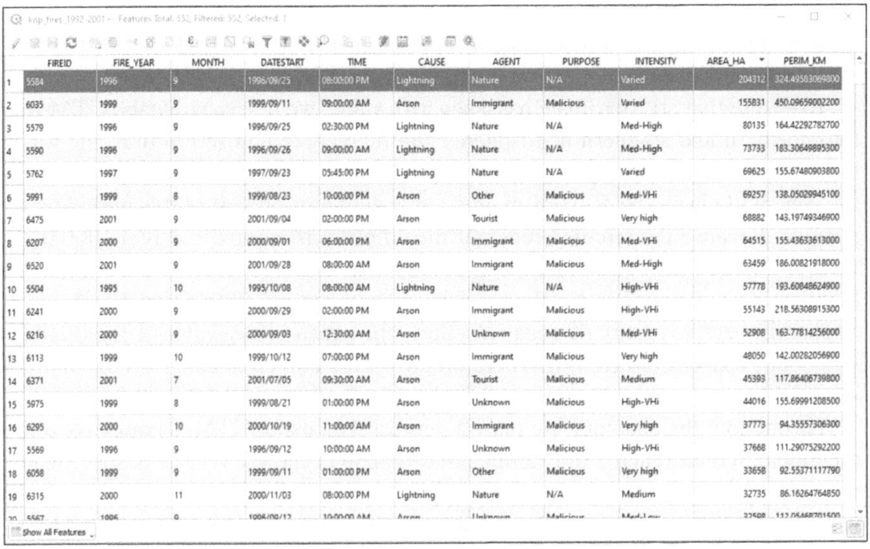

knp_fires_1992-2001 — Features Total: 552, Filtered: 552, Selected: 1

	FIREID	FIRE_YEAR	MONTH	DATESTART	TIME	CAUSE	AGENT	PURPOSE	INTENSITY	AREA_HA	PERIM_KM
1	5584	1996	9	1996/09/25	08:00:00 PM	Lightning	Nature	N/A	Varied	204312	324.49583069800
2	6035	1999	9	1999/09/11	09:00:00 AM	Arson	Immigrant	Malicious	Varied	155831	450.09659002200
3	5579	1996	9	1996/09/25	02:30:00 PM	Lightning	Nature	N/A	Med-High	80135	164.42292782700
4	5590	1996	9	1996/09/26	09:00:00 AM	Lightning	Nature	N/A	Med-High	73733	183.30649895300
5	5762	1997	9	1997/09/23	05:45:00 PM	Lightning	Nature	N/A	Varied	69625	155.67480903800
6	5991	1999	8	1999/08/23	10:00:00 PM	Arson	Other	Malicious	Med-VHi	69257	138.05029945100
7	6475	2001	9	2001/09/04	02:00:00 PM	Arson	Tourist	Malicious	Very high	68882	143.19749346900
8	6207	2000	9	2000/09/01	06:00:00 PM	Arson	Immigrant	Malicious	Med-VHi	64515	155.43633613000
9	6520	2001	9	2001/09/28	08:00:00 AM	Arson	Immigrant	Malicious	Med-High	63459	186.00821918000
10	5504	1995	10	1995/10/08	08:00:00 AM	Lightning	Nature	N/A	High-VHi	57778	193.60848624900
11	6241	2000	9	2000/09/29	02:00:00 PM	Arson	Immigrant	Malicious	High-VHi	55143	218.56308915300
12	6216	2000	9	2000/09/03	12:30:00 AM	Arson	Unknown	Malicious	Med-VHi	52908	163.77814256000
13	6113	1999	10	1999/10/12	07:00:00 PM	Arson	Immigrant	Malicious	Very high	48050	142.00282056900
14	6371	2001	7	2001/07/05	09:30:00 AM	Arson	Tourist	Malicious	Medium	45393	117.86406739800
15	5975	1999	8	1999/08/21	01:00:00 PM	Arson	Unknown	Malicious	High-VHi	44016	155.69991208500
16	6295	2000	10	2000/10/19	11:00:00 AM	Arson	Immigrant	Malicious	Very high	37773	94.35557306300
17	5569	1996	9	1996/09/12	10:00:00 AM	Arson	Unknown	Malicious	High-VHi	37668	111.29075292200
18	6058	1999	9	1999/09/11	01:00:00 PM	Arson	Other	Malicious	Very high	33658	92.55371117790
19	6315	2000	11	2000/11/03	08:00:00 PM	Lightning	Nature	N/A	Medium	32735	86.16264764850

Show All Features

FIGURE 4.7 Attribute table with the largest fire selected (highlighted in blue).

13. Save the format as an ESRI Shapefile.
14. Save the file in your working directory by clicking on … and name it 'Large_Fire'.
15. Ensure the CRS is the same as we have been working with in this chapter (EPSG: 32736).

Your dialogue box should look like that in Figure 4.8.

16. Click OK.
17. Turn off the full knp_fires_1992–2001 layer and the OSM basemap in the Layers Panel.

The extent of the fire should be the same as the outline in Figure 4.9. Given the importance of fires to livelihoods, one simple GIS analysis that we could undertake would be to measure the distance of the fire to the nearest towns. This would give an idea of the distance between the burned area and subsequent infrastructure and identify locations where preventative measures might be needed in the future. However, first we need to add more data to the project.

There are three common possibilities when adding datasets into our GIS projects.

- Adding a dataset with the same coordinate system
- Adding a dataset with a different projected coordinate system
- Adding a dataset with an unknown coordinate system

When we add a dataset with the same coordinate system, it opens without any warnings or messages, and aligns perfectly to where it should be. However, we should

Projections

FIGURE 4.8 Screenshot of parameters for save vector layer as….

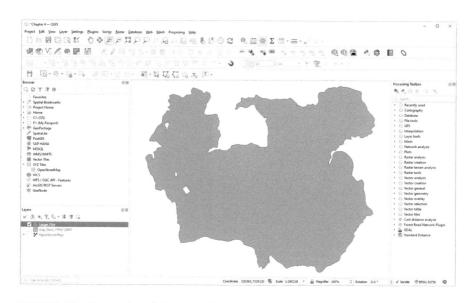

FIGURE 4.9 Screenshot of the largest fire represented as its own polygon layer.

always get into the habit of checking the projection when we import data as we may not always get a warning if a dataset is in a different projection.

When we add a dataset with a different projected coordinate system, sometimes we may receive notification, other times not. In the next step we add a layer representing the nearby towns to the large fire that occurred in 1996. This layer is projected in a different coordinate system, Sphere_Sinusoidal.

18. Navigate through the tab Layer > Add Layer > Add Vector Layer.
19. Add the Towns.shp.

We should see that the towns have been added to the map in the correct place. This is what we often refer to in GIS as 'projecting on the fly'. For visualization purposes, this can be good enough; however, should we wish to perform any analysis on the dataset, our analysis would most likely be impacted as the data is in a different projection. It is good practice to start projecting data into a uniform CRS right away, as errors and mistakes are less likely to impact our work.

20. Navigate through the tab Vector > Data Management Tools > Reproject Layer.
21. Select Towns as the Input layer.
22. Select the Target CRS as EPSG 32736 (hint: as we already have data in this projection within the map, it should be available within the dropdown options). The transformation should update itself.
23. In Reprojected, save the resultant layer permanently in your working directory as 'TownsUTM'.

Your dialogue box should look like that in Figure 4.10.

FIGURE 4.10 Screenshot of the parameters needed for the Reproject layer tool.

Projections

FIGURE 4.11 Warning notice for undefined coordinate reference system.

24. Click Run.

The new layer has been added, in the exact same location. This is what we expected, but if we open the properties for TownsUTM, we can see that the projection has been transformed. Finally, we are going to add a dataset with an unknown coordinate system.

25. Navigate through the tab Layer > Add Layer > Add Vector Layer.
26. Add the Airport.shp.

If the basemap is still visible and we have followed the steps, we should see that the data has aligned well. If the basemap was added first however (i.e., before the knp_ fires_1992–2001 layer), the data will be in the wrong location (most likely in the North Sea). Regardless, a warning is visible to the right of the layer in the Layer Panel, as shown in Figure 4.11.

This message means that QGIS cannot identify the CRS of the data. GIS software will try to display the data using the coordinate system of map, in our case it happens to be correct, but often it will be in the wrong location.

27. Click on the ? in the Layers Panel. This opens the Coordinate Reference Selector System. Select EPSG 32736 as the projection.
28. Click OK.

The ? should have disappeared, and we now have three layers all in UTM 36S. If we had zoomed-to layer, zoom out until we are once again viewing Kruger National Park. Now, we are going to undertake some simple measurement functions to identify the distance of the towns to the fire.

29. Open the attribute table for TownsUTM.
30. Select Makoko, and navigate to the town by clicking the Zoom to Selected Row(s). We may need to use the zoom button or mouse pad to get closer to the town.

You should be zoomed in to match the data in the interface presented in Figure 4.12.

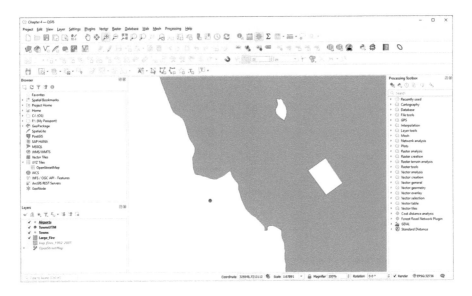

FIGURE 4.12 Screenshot of zoomed-in map canvas to the town Makoko.

FIGURE 4.13 Dialogue box of Measure Line tool.

31. Click on the Measure Line tool, this is the one that resembles a ruler.

A dialogue box should open that matches that of Figure 4.13. This should be an empty box linked to a tracker in the map canvas that will measure the total distance, along with two options: Cartesian and Ellipsoidal. Here, the GIS will calculate the distance along the ellipsoid or along the projected flat surface.

32. Click on the point representing the town, and then trace the line to the closest point on the fire polygon. We should see the distance increasing.
33. When we get to the edge of the fire polygon, click the mouse.

Projections

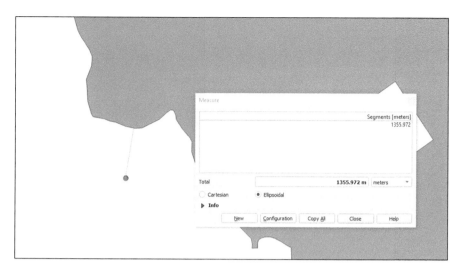

FIGURE 4.14 Screenshot of Measure Line tool in action.

The distance has updated and should read approximately 1.35 km or 1350 m, as shown in Figure 4.14.

34. Click on the Cartesian tickbox.

The distance changes by approximately 6 cm. It is quite difficult to note whether we select the exact same position if we repeat this analysis. To overcome this, we can turn snapping on.

35. Right click on the gray area to identify the toolbars that are hidden (as we explored in Chapter 2).
36. Turn on the Snapping Toolbar.
37. Click on the Magnet button, which will turn snapping on.

Now when we are measuring distances, it snaps to the exact location the feature is located at. If we repeat the analysis, the exact distance using this projection is 1367.019 m using Cartesian and 1367.024 m using Ellipsoidal.

The differences between the distances are because Cartesian distance is measured on a flat surface, while Ellipsoidal distance is measured on a curved surface. Ellipsoidal distance, also referred to as Geodetic distance in GIS nomenclature, is the shortest distance as measured by the defined ellipsoid. In this example, the distance is relatively small, so the measurements do not differ greatly. We could try measuring the distance from the north and south of Kruger National Park to see what the distortion is. Similarly, we could zoom out to the world view and measure the distance between South Africa and Australia, where we would see considerable differences between the two measurements. For a detailed discussion on the factors that cause these differences, see Sarkar (2019).

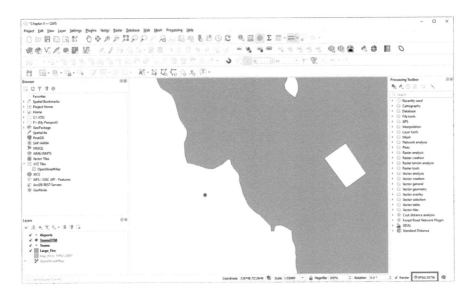

FIGURE 4.15 Screenshot of project coordinate reference system.

In QGIS, the coordinate information for the map canvas is provided in the bottom right of the screen, as highlighted in Figure 4.15. It is posted using an EPSG code, which stands for European Petroleum Survey Group, which is the organization that maintains a unique identifier for every projection system. This is useful as there are so many possible coordinate reference systems that the EPSG code makes sourcing these more manageable.

38. Click on the EPSG in the bottom right of the QGIS interface.

This should open another dialogue box as shown in Figure 4.16, which provides the information on the project we are working on, as well as any recently used CRS. Here, we can change the projection of the map frame and project. Click on the > under Predefined Coordinate Reference Systems to see just how many different projections there are.

39. Scroll down the Projected Coordinate Systems to World Mercator. Hint: we could always filter the options by typing this in the filter/search bar. There are three options, choose EPSG code 6893.
40. Select this projection, and click Apply.

Examine the data to determine how the Mercator projection distorts the map. We may want to click the Zoom Full tool to view the world in its entirety. Look at the relative sizes of Africa, North America, and Greenland. Africa is the second largest continent but with this projection North America and Greenland look as big (if not bigger) than Africa. Also note the extreme distortion around the poles.

Projections

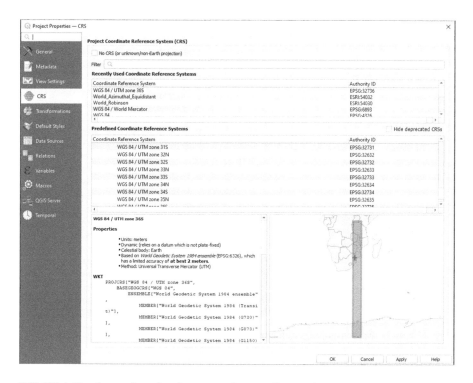

FIGURE 4.16 Screenshot of project properties coordinate reference system.

41. Repeat this process, but this time change the projection to World_Robinson. This has a code ESRI 54030.

The map is displayed using the Robinson projection. Examine the data to see how they have been distorted. This projection largely distorts all map properties, but minimizes angle variations at the equator, albeit with larger variations at the pole.

As mentioned earlier in this chapter, there is an almost endless list of projections that one could use. Spend a bit of time exploring different projections and seeing if you can ascertain the projection type and distortion. Some are obvious, others less so.

42. Once you have completed this, change the EPSG back to World Mercator (EPSG code 6893) and zoom back into Makoko.
43. Repeat the measurement using the Measure Line tool and with Cartesian ticked. The distance has now increased to 1510 m or 1.51 km.
44. Complete the same analysis for World Robinson. Here the distance is 1333.9 m or 1.33 km.

As we can see, these distances are beginning to differ more explicitly when we compare it to the Ellipsoidal distance. This is important if we consider such measurements might be used to support infrastructure development or exclusion zones in the event of a fire.

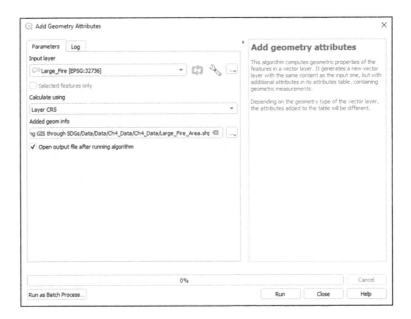

FIGURE 4.17   Screenshot of parameters needed for the add geometry attributes tool.

We can repeat similar analysis to investigate the difference the projections have on the area of the fire.

45. Navigate through the tab Vector > Geometry Tools > Add Geometry Attributes.
46. Select Large Fire as the Input Layer.
47. Select Layer CRS (which we know is UTM 36S) as the projection to calculate this value with.
48. Save the layer as a new file in your working directory, named Large Fire Area.

You dialogue box should look similar to that shown in Figure 4.17.

49. Click Run.
50. Open the attribute table of this new layer.

Two columns have been added to this dataset, representing the area and perimeter. These values match those already calculated in the dataset, which is not a surprise, but good to know that our result matches those already generated. Let us find out what happens if we repeat these steps with a different CRS.

51. Repeat steps 45–50, but instead of selecting Layer CRS in step 47, choose Project CRS, which should be World Mercator. If the map is not currently projected using this, do this prior to opening the tool. This projection should be saved in 'recently used layers' so should be quick to update.

World Mercator has estimated the size of this fire to be much larger, by approximately 45,000 ha in area and 20 km in perimeter. This is not surprising as the specific UTM zones were designed to overcome many of the limitations of calculating such statistics on a worldwide projection; however, the specific implementation of these steps in GIS software is not without a clear message regarding the huge impact that selecting appropriate projections can have.

4.3 CASE STUDY: CONCLUDING REMARKS

The choice of projection is particularly pertinent when we consider how many SDGs could be supported by such tools; area of land cover converted, area of land at threat from disaster, distance to vulnerable towns, or the area of vulnerable towns within such a threshold. Therefore, a basic understanding of what projections are, and why they are fundamental to GIS is imperative, particularly as we move through this book. Many of the examples in the analytical chapters will be based on linear units, which as we now know are highly influenced by the map projection. Therefore, when analyzing or presenting measurements derived within GIS, we must ensure we are selecting the most appropriate CRS for the task.

It is also important to stress that when working with datasets from multiple sources, oftentimes we can represent multiple layers in a GIS using different projections. Due to the ability of GIS software to project on the fly, for visualization purposes this is usually sufficient; however, we can run into problems if we begin to analyze this data. The first question I ask whenever myself or my students are having issues running any tools in GIS software is 'are all the layers in the same projection?' Often, they are not, and after you complete this book and begin your independent GIS career, I hope this chapter will resonate.

4.3.1 Test Yourself

If you want to test yourself on the learning objectives for this chapter, complete the following:

a. Identify the second largest fire over the ten-year period.
b. Calculate the difference in area using different projections for the second largest fire, including UTM zone 36S as well as others that we have explored within this chapter.
c. Identify the closest airport to the large fire from 1996.

REFERENCES

Marden, A.W., Meyer, T., and Crews Meyer, K.A., 2022. Regional fire occurrence in South Africa using BFAST iterative break detection in seasonal and trend components of a MODIS time series. *South African Geographical Journal*, pp. 1–22. DOI: https://doi.org/10.1080/03736245.2022.2066165

Martin, D.A., 2019. Linking fire and the United Nations sustainable development goals. *Science of the Total Environment*, 662, pp. 547–558.

Meyer, T., Holloway, P., Christiansen, T.B., Miller, J.A., D'Odorico, P. and Okin, G.S., 2019. An assessment of multiple drivers determining woody species composition and structure: A case study from the Kalahari, Botswana. *Land*, 8(8), p. 122.

Sarkar, D., 2019. *Distance Operations. The Geographic Information Science & Technology Body of Knowledge* (3rd Quarter 2019 Edition), J.P. Wilson (ed.). University Consortium for Geographic Information Science, Online. DOI: 10.22224/gistbok/2019.3.3.

Woods, D. and Govender, N., 2004. *SAFARI 2000 Historical Fire Maps, Kruger National Park, 1992–2001*. ORNL DAAC.

5 Attributes and Queries

5.1 INTRODUCTION AND LEARNING OUTCOMES

To successfully use GIS, we must understand how data is organized and what kinds of operations can be performed on these databases. We have already looked at some aspects of these skills in the previous chapters, where we have been introduced to the idea of navigating the attribute table, creating new data, and managing spatial data. In this chapter we advance our work on the attribute table to support the process of writing and performing queries, a fundamental GIS technique. By the end of this chapter, you will have completed four learning outcomes, and you should be able to:

- Combine spatial and attribute data using a table join
- Write and perform spatial and attribute queries
- Generate new data within the attribute table using the field calculator
- Perform simple geoprocessing tasks to manipulate and generate new data

In this chapter we analyze demographic and infrastructure data to support SDG9c. This target aims to increase access to information and communications technology and strives to provide universal access to the Internet. Here, we use queries to identify locations that could be targeted to support access in the Republic of Ireland using data compiled from the 2016 Census. While some of the data we use is already formatted for use in GIS, other data must be imported and linked to existing data before it can be used.

5.2 CASE STUDY: SDG9c TARGETED BROADBAND SUPPORT USING ATTRIBUTE QUERIES

1. Open QGIS and start a New Empty Project. If working chronologically through this book, you should now have two recent projects in your home screen.
2. Navigate through the tab Layer > Add Layer > Add Vector Layer.
3. Connect to the extracted Ch5_Data folder, and add the SAP shapefile.

Small Area Polygons (SAPs) are the smallest geographic unit that census data is compiled at in the Republic of Ireland, with this data obtained from the CSO via Ordnance Survey Ireland (OSi 2015) licensed under Creative Commons Attribution 4.0. One way of accessing the attributes for SAPs is to use the Identify Features tool. Clicking on an SAP with this tool will display a dialogue box called Identify Results showing all the attributes for that clicked tract (as shown in Figure 5.1).

4. Click the Identify Features tool and click on an SAP in the data frame.
5. Use the Identify Features tool to click on various SAPs and explore their attributes.

FIGURE 5.1 Screenshot of the small area polygon layer, with one polygon highlighted and linked to the Identify Features tool.

When we are zoomed out at full extent, sometimes multiple features are selected, highlighted in Figure 5.1 in bright red. Therefore, we may need to zoom in and explore an area in finer detail.

6. Save the project as Chapter 5. Remember it is good practice to do this throughout the process of working through the chapters.
7. Close the Identify Results dialogue box.
8. Right click on the SAP layer in the Layers Panel, and open the attribute table.

The attribute table is displayed showing a small number of attributes. Take a moment to familiarize yourselves with these attributes and work out what the various column names mean. A useful starting point with attribute tables is to identify how many records (i.e., rows) and attributes (i.e., columns) the table contains. This is important, as it provides the information on the number of spatial units (in our case records or features) and the amount of aspatial information (in our case the attributes). In this layer, we have 18,641 records (or features or rows) and 21 attributes (or columns).

Currently, while the table contains data relating to the SAPs, it does not contain any data about the overall population or broadband uptake. Luckily, we have data for broadband uptake for each of the SAPs in a separate text file, but this information is in tabular form and has no spatial properties itself. We must open this data and link it to the SAP layer's attribute table before we can map it. To link a nonspatial table with a layer attribute table, there must be a way to match records in one with appropriate records in the other. This is done with an attribute common to both tables, such as a name or identification code. A table join appends the attributes of the nonspatial table to the layer attribute table. The table join uses a key, with a primary key in the spatial layer and a secondary key in the aspatial layer. This key is unique to each record (or

Attributes and Queries

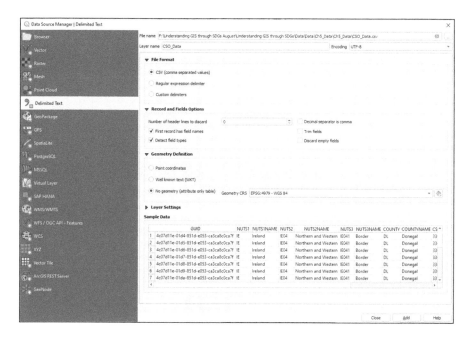

FIGURE 5.2 Screenshot of the parameters needed for the add delimited text layer function.

row or feature). The technical name given to a set of one or more attributes that can uniquely identify a record in a table is a ***unique identifier***.

9. Navigate through the tab Layer > Add Layer > Add Delimited Text Layer.
10. Browse to the location of the extracted folder, and select the CSO_Data.csv.
11. Under Geometry Definition, ensure No Geometry (attribute only table) is selected. This is not set as default, so we have to change this.

In newer versions of QGIS, we will be able to select the type of field (i.e., text or integer) for each attribute at this stage of the process, but in the current LTR this is not possible. We need all fields after T15_3_B to be reported as Integers, so we can work with the data accordingly. This will be built upon in more detail in the next chapter, but for now we can assume that QGIS has automatically assigned these fields correctly. Your dialogue box should resemble Figure 5.2.

12. Click Add (and close the dialogue box).

We should see that the table has been added above our spatial layer SAP, as shown in Figure 5.3. This data is sourced from the Central Statistics Offices (CSO 2018) licensed under Creative Commons Attribution 4.0. To use this data with the SAP layer, we need to link the two datasets together using a common field from both tables. This common field may have different names in each table (i.e., the title at the top of the column may be different) but the actual contents (i.e., the data held in each field) are the same.

FIGURE 5.3 Screenshot of the two layers in the Layers Panel, including CSO_Data that is saved only as a table, without any spatial information.

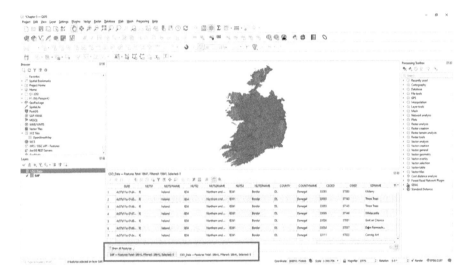

FIGURE 5.4 Screenshot of docked attribute tables, with the area highlighted in red the location to toggle between them.

13. Open the attribute table for SAP and for CSO.
14. Toggle between the two and try to identify the attribute or field that is common across both datasets. Hint: if we dock both tables, we can toggle between them easily and have both in semi-permanent view using the tabs highlighted in Figure 5.4.

There are several similar fields in both attribute tables that could fit the requirements of a unique identifier. Generally speaking, we want to try and avoid using names, particularly when it comes to census data gathered at a national level. Many towns within a country can have the same name, meaning we cannot be sure that the table will be joined to the correct feature. We can think of this as analogous to our own name. For example, your name is most likely not unique to you, especially when we consider this at a national or even international scale. Instead, we have other documentation that is unique, such as a passport number, social security

Attributes and Queries 67

number, student number, or staff number, among many other examples, and incidentally is one of the reasons we should not be sharing this information readily with people. This is the same logic we need to apply when performing table joins. For our Irish census data, the GUID is the unique identifier we need. However, when working independently in a GIS, we may not always be aware of what field is unique, meaning we could be tasked with searching the attribute tables ourselves, or even the metadata.

15. Right click on the SAP layer in the Layers Panel and open Properties. Alternatively, double clicking on the layer in the Layers Panel will also open Properties.
16. Navigate to the Joins tab.
17. Click on the green + sign toward the bottom left of the dialogue box.
18. Choose CSO_Data as the Join layer.
19. Next, we need to select the join and target field, which represent our primary and secondary key, or our unique identifier. We know this is GUID for both tables, so select accordingly.

Your dialogue box should resemble Figure 5.5.

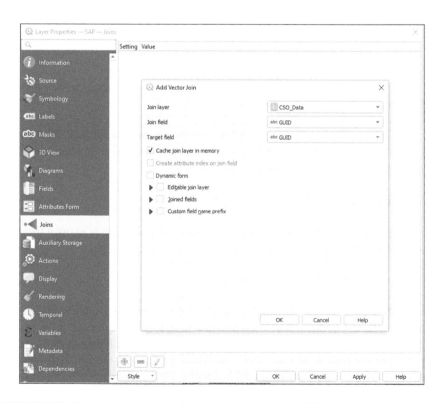

FIGURE 5.5 Screenshot of the join function in the properties dialogue box.

FIGURE 5.6 Screenshot of join in the properties dialogue box.

20. Click OK to assign the join.

If we click on the triangle, it outlines the parameters that we just specified for the join, as shown in Figure 5.6. For now, we do not need to add any further data to our project, so we can go ahead and complete the join.

21. Click Apply and close the dialogue box.

It may appear like nothing has changed. However, the next step is to open the attribute table for the SAP layer.

22. Open the attribute table for SAP layer.

The attribute table should look similar, but there should now be an additional five attributes (or columns) in this spatial layer. You may need to scroll to the end of the table. If the table contains several nulls, then the join has been specified incorrectly. Return to the previous steps and double check the parameters when setting the join. If this mistake has been made, we must remove or edit the existing join, which is achieved by returning to the properties dialogue box, selecting the join in question, and choosing either the red minus (–) button (or remove selected join) or pencil button (or edit selected join).

Attributes and Queries

Checking data after operations such as this is a very useful habit to get into. It can be easy to make a slight mistake or select the wrong unique identifier, which results in our data being incorrect or incomplete. By not checking our data, we risk propagating errors as the new data we are using could be incorrect. Throughout this book, we continue to implement these steps and check our data to ensure the data management and analytical processing we undertake runs correctly.

Returning to the case study, the CSO data is now linked to the SAP layer; however, this link is only temporary, and it is only preserved within this project. If we close QGIS without saving, the link will be broken. If we want to make the join permanent, such that we could use this data in another project, we need to export a copy of the file, using the same method we have implemented in previous chapters by right clicking on the layer in the Layers Panel. However, for now we can continue working with the temporary join, although this would be a good time to save the project.

Now that our layers are joined, we can select SAPs on the map and then look at their records in the attribute table. This can be very useful if we quickly want to compare certain features.

23. Click on the SAP layer in the Layers Panel to activate it, and then click on the Select Features button and select any SAP of your choosing.
24. Hold down the Shift key on the keyboard and click on an SAP adjacent to the first SAP selected. Because we were holding down the Shift key, both SAPs are selected.
25. Click the Deselect Features from all Layers button.

Interactive selection works when we can see exactly what we are looking for on the map (SAPs in a specific area, for instance). To select features according to qualities that we cannot see on the map (SAPs with several households over 100, for example) we write a query. A query selects features that meet specified conditions. The simplest query consists of an attribute (such as number of households), a value (such as 100), and a relationship between them (such as 'equal to' or 'greater than'). Complex queries can be created by connecting simple queries with terms like 'and' and 'or', using Boolean logic. Queries are not written in ordinary language (i.e., English) but in the database query language. All we need to do, however, is open a dialogue box and select the attributes, values, and operators we want/need. QGIS then creates the query.

26. Click on the Select Features by Value button.

A dialogue box should subsequently open, which has a list of all the attributes in this layer, as well as several options. In our newly joined layer, there is a variable called CSO_Data_T15_3_B. This is the number of households in the aggregated area that have broadband.

27. In the dialogue box, scroll down until we reach the attribute CSO_Data_T15_3_B.
28. Click on the Exclude Field button, which provides several relationships that can be specified. Choose the option of 'Greater than'.
29. Using the simple example provided earlier, type '100' in the text bar.

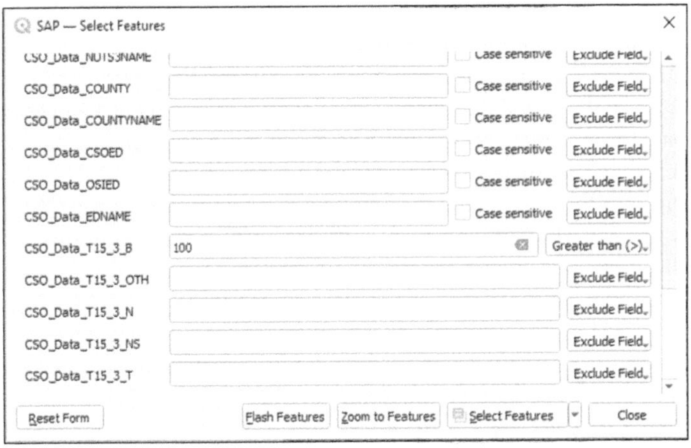

FIGURE 5.7 Screenshot of the parameters for the select features by values tool.

At the bottom of the dialogue box shown in Figure 5.7 there are a set of options, including Flash Features and Select Features.

30. Click Flash Features.

Assuming the dialogue box is not covering the entire SAP layer, we should have seen some features flash. If the dialogue box was covering the layer or you did not notice it, try it again. Flashing features is useful if we want to quickly know how many features match the query, but it does not allow for any detailed interrogation of the data. To make this selection less temporary we use select features.

31. Click Select Features.

All the features (or SAPs) that have over 100 households with broadband are selected in yellow, as shown in Figure 5.8. A blue banner should also appear at the top of the map stating that there were 1404 features selected. This is a semi-permanent selection. In some instances, we may wish to make this permanent by exporting the data, and again, we revisit this in future chapters. For now, let us create some more queries.

32. Clear all the selected features by clicking the Deselect Features from all Layers button.

Next, we will compare two attributes in the feature layer. With the aim of SDG9c to identify universal access to broadband, it would be useful to identify locations where the number of houses with broadband is less than the number of houses in that SAP.

33. Open Select Features by Values if it is not already open and click Reset Form.

Attributes and Queries 71

FIGURE 5.8 Screenshot of the QGIS project after select features has been clicked. There are selected SAPs in yellow, and the blue banner stating there are 1404 features selected.

34. Scroll down to CSO_Data_T15_3_B.
35. For this query we want to choose the operator 'Less Than'.
36. In the text box, instead of typing a number, type the name of the attribute that represents the number of households. For reference this is CSO_Data_T15_3_T, and your dialogue box should resemble Figure 5.9.
37. Click Select Features.

It should look like the whole country has been selected; however, we should always aim to be more precise.

38. Open the attribute table for the SAP layer.

The attribute table should have almost all the records/rows highlighted. The number of selected features should be reported at the top of the attribute table. We should see that 18,629 out of 18,641 have been selected. Perhaps a better way of asking this question might have been to choose 'Equal to' as the operator. However, rather than start the process again, we can invert our selection in the attribute table.

39. Click the Invert Selection button.

Our selection has inverted, but with so many features it is difficult to see exactly which SAPs fulfill this requirement.

40. Click Show All Features at the bottom of the attribute table, and select Show Selected Features, as shown in Figure 5.10.

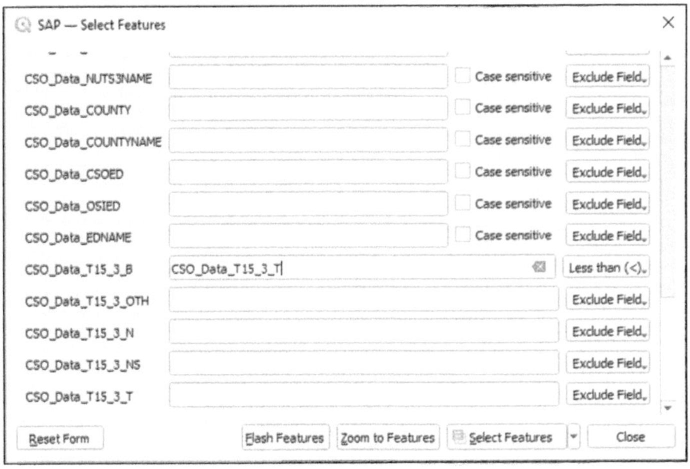

![Figure 5.9 screenshot]

FIGURE 5.9 Screenshot of the query to identify locations where the number of houses with broadband is less than the number of houses in total.

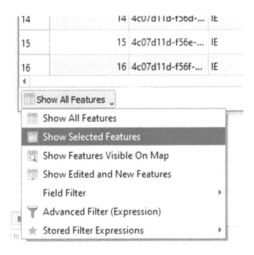

FIGURE 5.10 Location of show selected features in attribute table.

This should return only the 12 features that match the inverted criteria. We may want to see where these are on the map.

41. Click on the Zoom Map to Selected Rows button in the attribute toolbar.

Your map should resemble Figure 5.11. All 12 locations are within the greater Dublin region. This is the capital of the country, meaning that as of 2016, there are no locations outside of the capital where every household has broadband. Of course, in many locations, the installation of broadband is optional on the homeowner or tenant rather

Attributes and Queries

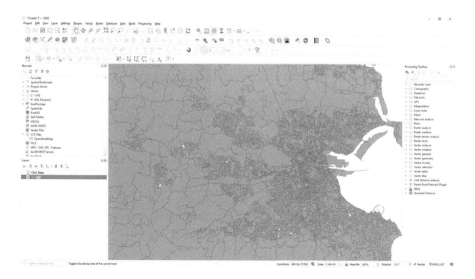

FIGURE 5.11 Screenshot of the selected features that match the query whereby the number of houses with broadband matches the total number of houses in the SAP. For reference, this is the greater Dublin area.

than it not being available. However, if a large proportion of houses in the area do not have broadband, then this could be indicative of a connectivity issue. Therefore, it would be useful to have a value that represents the percentage of an SAP that has broadband. To do this, we add a new field to our attribute table.

42. Click the Deselect Features from all Layers button to remove the current selection.
43. Select 'Show All Features' in the attribute table if it looks blank (see step 40).
44. Click on the Toggle Editing button in the attribute table.

This function allows us to edit the attribute table, such as changing values or adding and/or deleting attributes. Therefore, take care when working with Toggle Editing switched on so as not to change any of the values by mistake.

45. Click on the New Field button.
46. Name the new field 'BB_Percent'.
47. Add a comment of 'percentage of households with broadband'.

By commenting on new data such as this, it supports our re-use of the data in the future in case we have forgotten exactly what the information is. This also supports other users to work with our data and is good practice for any aspiring GIS specialists.

48. Change the Type from Integer to Decimal Number.

Your dialogue box should resemble Figure 5.12.

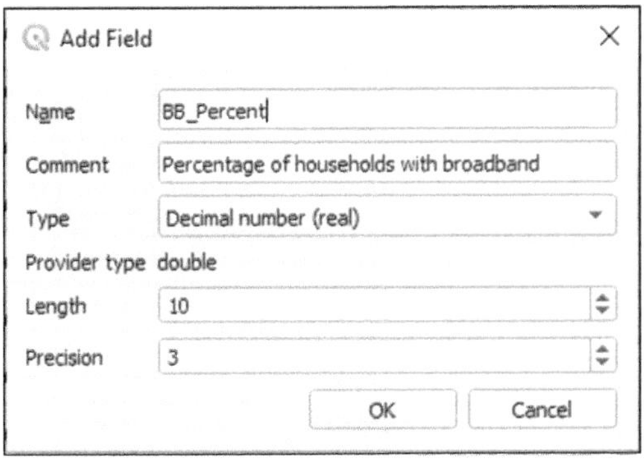

FIGURE 5.12 Parameters for the add field tool in the attribute table.

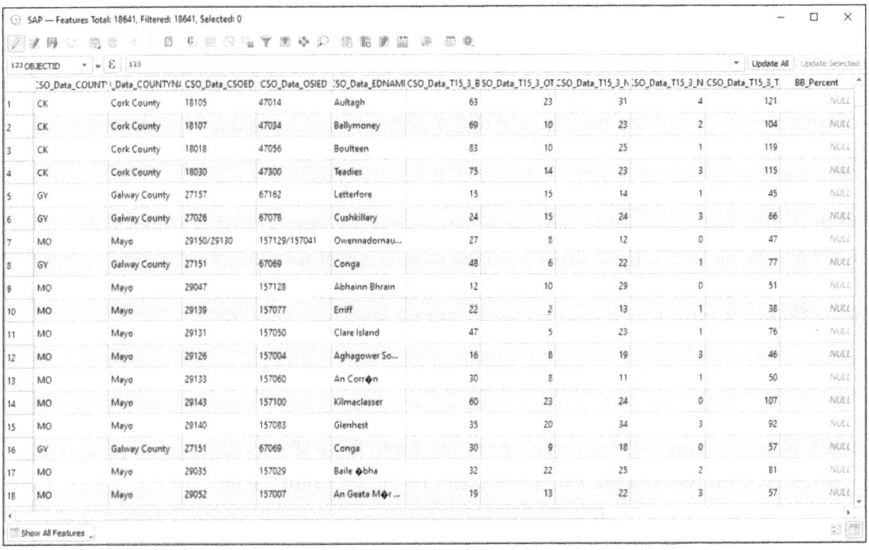

FIGURE 5.13 Attribute table with NULL values in the newly created BB_Percent field.

49. Click OK.

This may take a minute to update. Remember, it is a large vector file.

50. Scroll across to the end of the attribute table.

We should see a field/column/attribute called BB_Percent, with all the values specified to NULL as shown in Figure 5.13.

51. Click on Toggle Editing again and save our work. This makes the new field permanent, meaning we cannot accidently change any of the values.

Attributes and Queries

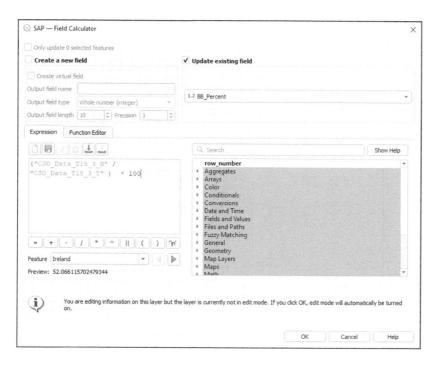

FIGURE 5.14 Screenshot of the parameters and expression needed for the field calculator to capture percentage of households with broadband in the small area polygon layer.

52. Open the Field Calculator tool by clicking on the button in the attribute table.

Another dialogue box should open. The default setting is to create a new field. However, we just completed those steps, meaning we want to update an existing field instead.

53. Check 'Update Existing Field' and choose BB_Percent.
54. Using the Expression tab, we want to type in the following expression: ("CSO_Data_T15_3_B" / "CSO_Data_T15_3_T") * 100.

Your dialogue box should resemble Figure 5.14. Ensure that this expression contains the quotation marks. This is how the field calculator specifies this text as a field (or attribute). In the bottom left, there is a preview option. This states that based on the expression, the first feature in the attribute table will have a value of 52.066. This is within the bounds of possible (i.e., 52.066%), meaning our expression looks like it is correctly formatted. We can check different features to ensure that our values do not extend beyond the correct range of 0% and 100%. It should also be noted here that if the expression is incorrect, we get a warning message that the expression is invalid. If this occurs, the field calculator will not let us complete the update, with the OK button grayed out. In the below example in Figure 5.15, there is a quotation mark

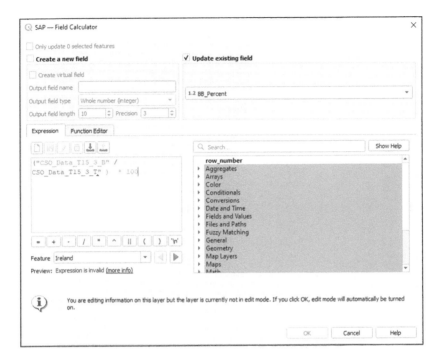

FIGURE 5.15 Screenshot of incorrect expression in the field calculator that results in an invalid expression and a failure to complete the function.

missing from one of the attributes. Notice the different color used in the Expression to denote text as opposed to a field/attribute name, as well as the warning message and grayed out button.

55. Ensuring the expression is correct (i.e., Figure 5.14), click OK.

Note, this may take a little while to update. Most times I have processed this it has taken between 30 and 60 seconds. Once this has finished processing, we should see that the column has been populated with percentage values.

56. Click on the column heading BB_Percent to sort the values. Clicking it once sorts it in ascending order, while clicking it twice sorts it in descending order.

We should see from sorting the values in BB_Percent that the lowest broadband percentage in Ireland is 2.5%. We know that the attribute data is linked with the spatial data, so we can zoom to this location.

57. Select the row/record in the attribute so that it is highlighted in blue.
58. Click the Zoom map to Selected Rows button in the attribute table.

You should be zoomed in so that your map resembles Figure 5.16.

Attributes and Queries

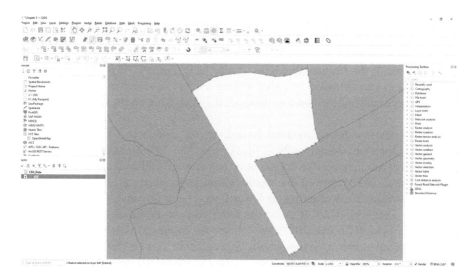

FIGURE 5.16 Screenshot of the SAP with the lowest broadband percentage in the Republic of Ireland in 2016.

Note the red crosses around the edge of the polygon. These are the basic spatial primitives of the polygon, points or vertices that are connected by straight lines as explained in Chapter 3.

We have now found the lowest area of broadband connectivity, but we must ensure we interrogate the results of spatial analysis alongside geographic context, which is easily found in the form of basemaps or reference maps.

59. In the Browser Panel, navigate to XYZ Tiles and turn on the OpenStreetMap reference layer.

This will provide us with a spatial reference for our data, which should overlay the SAP layer.

60. Rearrange the layers in the Layers Panel so that SAP is above the OSM reference layer.

The area is a small city block that contains university student accommodation. We need to think critically about why this area might have such a low percentage (<3%), and whether this is a true representation of the processes operating in the area, or an artifact of the data collection methodology. As extreme values can oftentimes be outliers, it would be useful to select multiple features that are below a specific threshold, such as less than 20% of the households. To do this, we write another query.

However, as we started a toggle editing session, we should not perform any more functions until these edits are first saved. In fact, QGIS will prevent us from completing specific functions until this occurs.

61. Click on the Current Edits button in the Digitizing Toolbar and select save for the selected layer. This will make the addition of the new field and values permanent.
62. Select all the features that have a value of less than 20% for BB_Percent. This can be achieved by either writing a query or selecting multiple attributes in the table by holding down shift and clicking on the records.

We know that the lowest location is in Limerick, which is a city; however, the remaining locations with broadband uptake less than 20% are all predominantly in rural areas (you may need to zoom out or pan around), although there are also two SAPs in Cork and Waterford that would be considered urban areas. It is well established that rural areas often report lower Internet and broadband adoption (Mora-Rivera & García-Mora 2021), as well as lower availability, speed, quality of service, and higher price (Grubesic 2017). We can use the basemap to explore whether there might be any patterns in the topography or geographic location that might explain the low uptake. Generally, these locations are spread around the country, but there does appear to be a trend toward coastal areas or locations adjacent to large water bodies, suggesting that lack of infrastructure may be the reason for this low percentage. This spatial information is important for decision-makers to consider when identifying strategies to support SDG9c.

Another aspect of broadband accessibility is in relation to education, ensuring that areas close to schools have access. This is particularly pertinent given the shift in emphasis on home-learning since the pandemic, and the need for infrastructure to support education (Ruiz-Martínez & Esparcia 2020). To identify the number of schools within an area of low broadband uptake, we use spatial queries.

Firstly, we must export the selected features to create a new spatial layer.

63. Ensuring the selection is still in place, right click on the SAP layer and choose Export Data > Save Selected Features As.
64. Save this as a shapefile in your working directory as SAP_BBLow.

Your dialogue box should resemble Figure 5.17.

Now would also be a good time to save your project if you have not done so in a while.

65. Navigate to the tab Layer > Add Layer > Add Vector Layer.
66. Add the school shapefile from the extracted folder.

This should insert the point layer of all post-primary schools in Ireland. This is the same layer we used in Chapter 3, from the All-Island Research Observatory (AIRO 2016) licensed under Creative Commons Attribution 4.0.

67. Click on the Select by Location button in the Selection Toolbar.
68. Select Schools as the Select Features from.
69. Select the geometric predicate as Intersect.
70. Compare this to the newly created layer SAP_BBLow.

Attributes and Queries

FIGURE 5.17 Parameters to save selected features as a new shapefile.

Your dialogue box should resemble Figure 5.18.

71. Click Run.

None of the schools have been selected, but some are very close. As SAPs, like most census tracts, have somewhat arbitrary geographic boundaries (known in geography as the modifiable areal unit problem), it is useful to consider distance in such a query. To identify all schools within a specified distance of an SAP with low broadband uptake, we must use a buffer to create a new layer that considers distance.

A quick comment on the Select by Location dialogue box. There are various topological relationships that could have been used as the geometric predicate. Topology is the study of geometrical properties and spatial relationships that are unaffected by distance. For example, identifying two feature layers that intersect, touch boundaries, or contain each other provides information that may be lost when distance is used. This is particularly relevant given our experiences in the previous chapter where we know that if we change the underlying projection, we may get different distances. In this instance, there is no topological relationship that can be used, meaning we must consider distance; however, we use distance in the knowledge that the underlying coordinate reference system may impact our results. This is a common feature of GIS work and part of the reasoning projections were introduced so early in this book.

72. Navigate through the tab Vector > Geoprocessing > Buffer.
73. Select SAP_BBLow as the Input Layer.

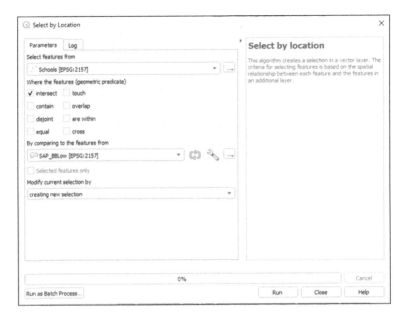

FIGURE 5.18 Parameters for the Select by Location tool.

We should see that QGIS reports the EPSG code for the Input layer, again subtly reinforcing the need to consider and appreciate projections in these geoprocessing operations.

74. Change the distance to 1 km.
75. Save the output as SAP_BBLow_Buffer as a shapefile in your working directory.

Your dialogue box should resemble Figure 5.19.

76. Click Run.

There should now be a new layer that looks similar to the SAP_BBLow layer, but with a 1 km buffer around it. Buffer is simply a geoprocessing tool that generates a new feature layer with a pre-defined radius or neighborhood surrounding it, in this instance 1 km. We return to buffers and neighborhoods in much more detail in Chapter 11.

77. Repeat the steps to select schools that intersect with SAPs with low broadband, this time using SAP_BBLow_Buffer as the 'by comparing to the features from' option.

We now have information on schools that are within a 1 km catchment of a low broadband uptake SAP. When considering SDG9c alongside SDG4 Quality

FIGURE 5.19 Parameters required for the buffer tool.

Education, these are areas that warrant further research to investigate whether targeted funding or infrastructure could support increased broadband uptake/accessibility at home to support education. For reference, we should have 17 schools selected.

5.3 CASE STUDY: CONCLUDING REMARKS

We have reached the end of the third case study chapter, continuing our data management and implementing queries within GIS. In this chapter we have learnt how to combine spatial and attribute data using a table join, highlighting the importance of the unique identifier. We generated new data in the attribute table, using the field calculator tool, which is a fundamental tool that will be prevalent throughout the remainder of this book (and any subsequent GIS careers). We also demonstrated the ability of GIS to perform queries and use these to immediately identify locations that have a lower uptake of broadband at a national level, which might point to areas that need attention from policy makers to support SDG9c, particularly in rural locations of Ireland that might be lacking infrastructure. Finally, we performed spatial queries, where we query the data using topological and distance relationships with other variables, in this instance schools. In achieving this, we undertook simple geoprocessing tasks, such as buffer, to manipulate existing data and generate new data. This final task demonstrated how to implement spatial queries in GIS, but also the need to consider multiple SDGs in our analytical framework to maximize the impact that GIS can make in supporting these targets, in this instance SDG9c and SDG4a.

5.3.1 TEST YOURSELF

If you want to test yourself on the learning outcomes of this chapter, complete the following:

a. Identify the areas with the lowest 'other' form of Internet connection (T15_3_OTH). This includes use of mobile data, such as 3G and 4G. Explore these results to see if these areas overlap with areas of low broadband. You can formally test this using a spatial query.

REFERENCES

Grubesic, T.H., 2017. Future shock: Telecommunications technology and infrastructure in regional research. In *Regional Research Frontiers*, Vol. 1 (pp. 51–70). Springer, Cham.

Mora-Rivera, J. and García-Mora, F., 2021. Internet access and poverty reduction: Evidence from rural and urban Mexico. *Telecommunications Policy*, 45(2), p. 102076.

Ruiz-Martínez, I. and Esparcia, J., 2020. Internet access in rural areas: Brake or stimulus as post-Covid-19 opportunity? *Sustainability*, 12(22), p. 9619.

6 Data Management

6.1 INTRODUCTION AND LEARNING OUTCOMES

Managing our data in GIS is perhaps one of the most important practical skills we can master. However, there is no surefire recipe for success when it comes to data management, and everyone will have their own preferences as to how best to achieve a working harmony in a GIS environment. In this chapter, we explore components of practical GIS work pertaining to data format, temporary versus permanent layers, attribute formats, and vector geometries. These are a selection of data management functions, tools, and tips that continually arise each semester during my GIS teaching. As such, this chapter contains specific instructions to complete the remainder of the book in an efficient format, but also content that will support independent GIS work upon completion. By the end of this chapter, you will be able to:

- Select between vector data formats and save layers within a GeoPackage
- Discuss the differences between temporary and permanent layers
- Open a QGIS project file and fix broken links
- Check and fix geometries and spatial indices in vector data

The data used in this chapter draw on that used in other chapters. The CSO_data from Chapter 5 was sourced from the Central Statistics Offices (CSO 2018) licensed under Creative Commons Attribution 4.0. Global_Landslides is a point file of global landslide locations from Kirschbaum et al. (2010, 2015) downloaded via Data.gov licensed under Open Data Commons Open Database License (ODbL) v1.0. SoilType is a polygon representation of soil type converted from the WISE30sec soil properties global grid (Batjes 2016) licensed under CC-BY-3.0. The Vacant and Derelict Land Survey (Buildings), Public CCTV locations (CCTV), and the Edinburgh Ward Boundaries (CityBoundary) were all sourced from the City of Edinburgh Open Data Portal (https://data.edinburghcouncilmaps.info/). These datasets are licensed under the Open Government License v3.0. These datasets are Copyright City of Edinburgh Council, contains Ordnance Survey data © Crown copyright and database right (2022). The Roads layer is the HOTOSM UK Scotland Roads dataset, sourced from the Humanitarian Data Exchange Portal licensed under the Open Database License (ODC-ODbL), and clipped to the Edinburgh area. The train station and tourist sites were digitized by the author.

6.2 VECTOR FORMATS

There are various vector storage formats that are used within GIS software. In the previous chapters, we exported or saved the selected features as a new layer, and in the QGIS dialogue box, we had the option to select from over 20 different formats,

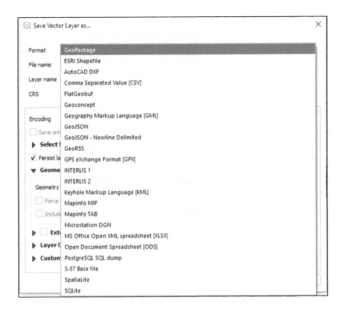

FIGURE 6.1   File format that vector layers could be saved as in QGIS.

listed in Figure 6.1. Up to this point in the book, we have used shapefiles to store our spatial data. Shapefiles are a simple storage format that were developed in the early 1990s.

A shapefile is a file format for storing the geographic location and attribute information of geospatial features. As outlined in Chapter 3, shapefiles can be points, lines, and polygons, but there are multiple files that make up the shapefile. The three mandatory files are *.shp, *.shx, and *.dbf, and *.prj is the projection file that defines coordinate reference system. In Chapter 4, you may have noticed that the *.prj file was missing from the airport layer, and subsequently why we needed to define that projection. While the term shapefile relates to the *.shp file, this file alone is insufficient for the file to be read in a GIS. You may also come across other extensions affiliated with shapefiles, meaning sometimes we can have up to seven to eight extensions for the same spatial layer. Therefore, it is important when saving or transferring shapefiles that we move all the files associated with it.

Within QGIS, there is a built-in panel for managing our spatial data, the Browser Panel.

1. Open a New Empty Project in QGIS.
2. Navigate to the Browser Panel. Remember if it is turned off, right click on the gray toolbar area (see Chapter 2 for instructions), and turn it on.
3. In the Browser Panel, navigate through your folders to the file directory where Ch6_Data has been saved and exported.
4. Click on the triangle next to the folder Ch6_Data_Part_1 to expand it.

Data Management

FIGURE 6.2 Screenshot of the Browser Panel.

We should see several shapefiles within the folder, as shown in Figure 6.2. This looks very different to how we have observed these files in the folders to date, as the Browser Panel has provided only one file per spatial layer, despite us knowing that there are in fact at least four digital files associated with the shapefile.

5. Right click on Buildings.shp and select 'Show in Files'.

This will open the file window with the raw files where this difference can be observed directly.

6. Navigate back to QGIS and the Browser Panel, right click on Buildings.shp and select 'Layer Properties'.

This should open another dialogue box that outlines the properties of this specific layer, with three tabs, Metadata, Preview, and Attributes, as shown in Figure 6.3. Metadata should be the default, and this should contain all the data on this data layer, including its location, projection, etc. Preview provides a spatial representation of the data, while Attributes provides non-spatial attribute data affiliated with this file; these are shown in Figure 6.4.

7. Explore the Metadata, Preview, and Attributes tabs for Buildings.
8. After sufficient exploration, close this dialogue box.
9. Double Click on the Buildings.shp layer.

This should add the data to the map. Adding layers directly from the Browser Panel can be preferable to the specific tool that we have been using thus far. Throughout the book, we continue to navigate through the tab Layer > Add Layer as this provides us

86   Understanding GIS through Sustainable Development Goals

FIGURE 6.3 Screenshot of layer properties dialogue box.

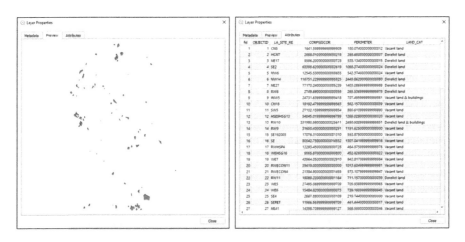

FIGURE 6.4 Screenshot of the layer properties dialogue box for preview where we can see the spatial dimension of the data and attributes where we can see the attribute table.

Data Management

with more control over the features that are being added. However, do note that the Browser Panel gives an in-built data management system that can handle efficiently the nuances of spatial data.

10. Turn on the OpenStreetMap XYZ tile to observe where this data is.

We should be looking at derelict buildings in Edinburgh, although the project projection is not orientated intuitively.

11. Change the project projection to British National Grid EPSG 27700 using the CRS option in the bottom right of the project. Click OK for any subsequent transformations that are needed.

This should have re-orientated us, using an on-the-fly projection. We return to Edinburgh in Chapter 18 where we implement route selection. For now, we are interested in how we manage spatial datasets. Despite the widespread use of shapefiles, they do have limitations. Shapefiles can only store information on one feature, and with multiple files associated with them they can be difficult to manage within a folder setting. They cannot store attribute names that are longer than ten characters, nor store data with both time and date formats, which can be problematic for data where this is important (more on that in Chapter 10). Subsequently, dedicated spatial data management systems have been developed to specifically store spatial data.

A GeoPackage in QGIS is a standard OGC data package that allows us to store various types of spatial data. This is the main advantage of GeoPackages in QGIS, as it provides a single location with which to save our data. These are not compressed like their counterparts in other GIS software, which means that the file sizes can get large; however, there are methods to compress these upon completion of geoprocessing and data manipulation. To support data management using GeoPackages, we can convert our shapefiles.

12. Add the remaining layers from Ch6_Data_Part1 using the Browser Panel.
13. In the Processing Toolbox, search for Package Layers. It should be within the Database options. Open this tool.
14. Under Input Layers, click on the ... and select all the layers. Click OK.
15. Save the destination package in your working directory as Ch6_DataP1_GeoPackage.
16. Ensure your dialogue box resembles Figure 6.5 (with six input layers selected). Click Run.

All these layers are saved within the new GeoPackage. Obviously moving forward, we do not need to specify that this is a GeoPackage in the name, it should be evident from the file extension, but to prevent any confusion in interpretation at this point we have added it.

If we expand the folder in the Browser Panel, as shown in Figure 6.6, we can see each of these layers. We can add each of these layers manually to our project using

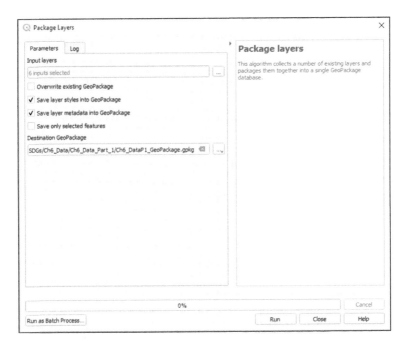

FIGURE 6.5 Screenshot of parameters for package layers tool.

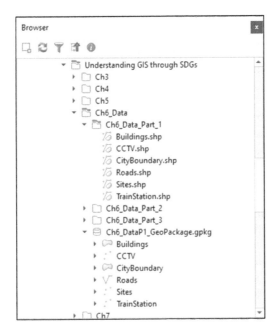

FIGURE 6.6 Screenshot of Browser Panel with new GeoPackage.

Data Management 89

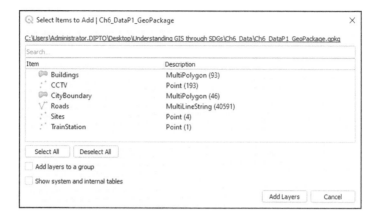

FIGURE 6.7   Screenshot of select items to add within the GeoPackage.

the method outlined above, but for now we will use the Add Layer function we have used so far in this book to demonstrate how to add data from a GeoPackage.

17. Navigate through the tab Layer > Add Layer > Add Vector Layer.
18. Navigate to your working directory. There should only be one *.gpkg file visible. This is expected.
19. Click Open to select it, and then Click Add.

This time an additional dialogue box should appear highlighting all the layers that are stored within the GeoPackage. This is shown in Figure 6.7.

The brackets after each description illustrate how many features there are. In this instance, there is one point representing the train station, and 40,591 lines representing roads.

20. Click Add Layers.

This will open all layers within this GeoPackage. If we only want one layer from the GeoPackage, we select this when adding the layers. There should also be a distinction between layers in a GeoPackage containing the name of the GeoPackage and the shapefiles, as shown in Figure 6.8. This will allow us to distinguish between them if needed.

At this stage in our GIS careers, it would also be useful to learn how to save spatial data inside an existing GeoPackage.

21. Ensure the Buildings layer from the Ch6_DataP1_GeoPackage is highlighted in the Layers Panel, and click on the Select by Values button.
22. We want to select all derelict buildings. In the dialogue box, specify the query that Land_Cat is equal to derelict buildings, as shown in Figure 6.9.
23. Click Select Features.

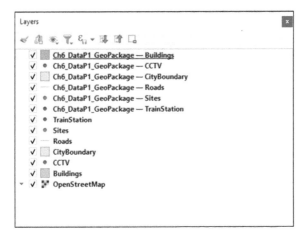

FIGURE 6.8 Screenshot of the layers panel. Note the difference in naming convention between the shapefiles and layers within a GeoPackage.

FIGURE 6.9 Screenshot of the select features query to return all derelict buildings.

There should only be one building selected, which we want to make permanent.

24. Right click on Ch6_DataP1_GeoPackage – Buildings in the Layers Panel, and select Export > Save Selected Features As.

A new dialogue box should open. The first item we need to select is called Format. If we click on the dropdown menu, we can see the 20 different data formats identified at the start of this chapter in Figure 6.1. There are several different data formats we could work with in our GIS journey, but for now we focus on GeoPackage as the predominant format.

25. Ensure GeoPackage is highlighted.

File name specifies the GeoPackage location and file. So, if we click on the ... button, we should be taken to our file manager, and we can navigate as usual to different sub-folders.

26. Click on the ... button of file name and navigate to the GeoPackage that was created a moment ago, called Ch6_DataP1_GeoPackage.

Data Management

Here we have two options. We can type directly into the file name bar at the bottom of the dialogue box. This will create a new GeoPackage. Alternatively, we can double click on Ch6_DataP1_GeoPackage.gpkg to select that as the location we wish to save our new layer in. We do not need to create another GeoPackage, so we select the existing one.

27. Double click on the *.gpkg file to select it.

This should take us back to the Save Vector Layer As … dialogue box. In the third bar, Layer Name, we specify what we want this layer to be called.

28. In Layer name, type Derelict_Buildings.
29. Specify that the CRS should be British National Grid EPSG 27700.
30. Finally, under Layer Options, ensure that Spatial Index is ticked (more on this later in the chapter).

There are options to add further information, such as data source. It is always good practice to try and complete this information, if just for our own benefit when we return to data after a period away. When we complete these steps throughout the book, I do not highlight this as a step, but always try remembering to populate this if you can. Now that we have completed the dialogue box, we can finalize the layer creation.

31. Ensure your dialogue box resembles Figure 6.10. Click OK.

This creates a new layer in the existing GeoPackage, which should appear in the Layers Panel.

When we navigate to the *.gpkg in our file management system outside of QGIS, we can see that there is just one file affiliated with it. This can support effective data management and file sharing. From this point in the book, we use GeoPackages to store, create, and share our spatial data. Therefore, when exporting data to the working directory or GeoPackage, please use these steps. For each chapter, try saving all newly created layers in one GeoPackage to reduce the amount of data files that are being generated.

Please note, there are some tools that do not provide the option to save as a GeoPackage or as a layer within one. As such we have to save as another format. In these instances, I suggest simply using shapefiles, and I will highlight where this is necessary as we progress through the book.

Another recurring aspect of saving layers within QGIS through teaching and classroom testing of this book is the process of overwriting layers. QGIS offers the functionality to overwrite layers, but as with any software this should come with warnings. QGIS will provide a warning that we are about to overwrite an existing file, so we have the opportunity to retract the command if requested. However, in my experience, and that of colleagues and students, this is not always straightforward and may not always work, especially if the layer is still within the QGIS project. Therefore, take caution when using the overwrite function, but also be aware that

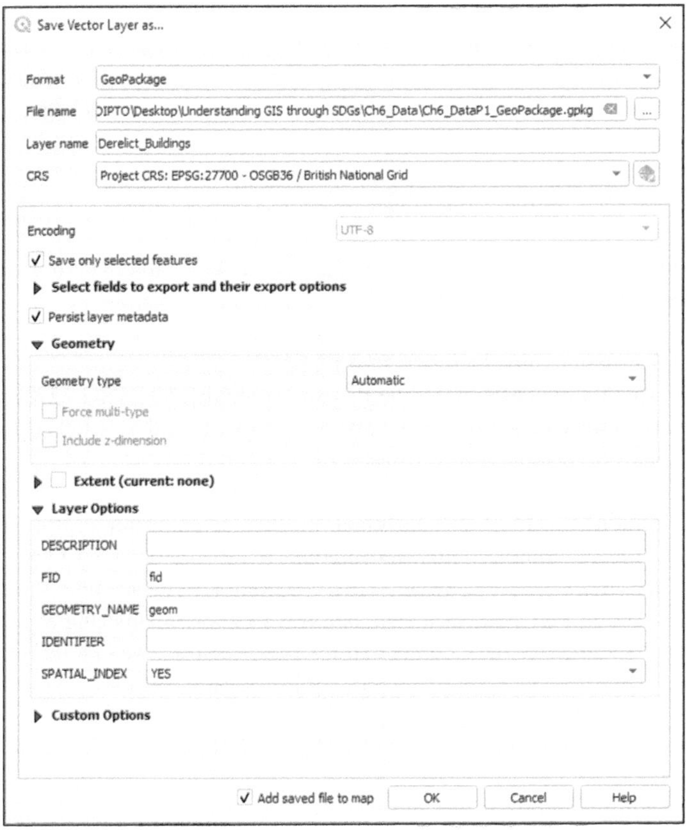

FIGURE 6.10 Screenshot of parameters for saving layers in an already existing GeoPackage.

this might not operate as smoothly as it should, and we may be safer creating a new feature layer or a temporary layer.

Finally, the only other layer format we have not yet discussed is rasters. Rasters tend to be more straightforward within a GIS, although there are several formats that can be used. To ensure learning outcomes are met with regard to the practical work, we work with rasters in *.tif format only. While there are other data formats available, in my experience this tends to be the best format to share data across platforms, as well as reducing data volume. Upon completion of this book, I would suggest exploring different formats of rasters, but the analytical components that are afforded within them are broadly the same.

6.3   TEMPORARY LAYERS AND BROKEN LINKS

We may have noticed at this point in the book that all the tools provide an option to save temporary layers instead of writing these to the hard drive. This can be useful if experimenting with the tools and not committed to parameterization; however, it

Data Management

can also create a very confusing Layers Panel if used multiple times as all layers are given the same name.

32. Navigate through the tab Vector > Geoprocessing Tools > Buffer.
33. Select Derelict_Buildings as the Input.

For now, we can keep all values as default. In the Buffered option, there is a grayed-out area called [Create Temporary Layer]. If this is left blank it will create a temporary layer, while if we click on the … and save this to our GeoPackage it will save it as a permanent layer. Let us run the tool creating a temporary layer.

34. Ensure your dialogue box resembles Figure 6.11. Click Run.

In the Layers Panel we now have a layer called 'Buffered'. If we repeat this process using a different distance, we get another temporary layer called 'Buffered'.

35. Repeat the buffer for Derelict_Buildings at 100m, creating another temporary layer.

Your Layers Panel should resemble Figure 6.12. Not only is this confusing, but as is denoted by the temporary symbol on the right of the Layers Panel, should we close

FIGURE 6.11 Screenshot of parameters for the buffer tool.

FIGURE 6.12 Screenshot of layers panel with two temporary layers named 'buffered'.

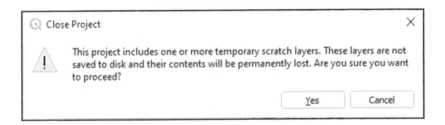

FIGURE 6.13 Screenshot of warning that the project contains temporary layers.

QGIS, these layers will not be saved. Therefore, temporary layers are highly useful during the exploration phase of GIS, but we should aim to not let our project contain too many layers of a temporary nature.

36. Save the Project as Ch6 and close QGIS.

The warning statement shown in Figure 6.13 appears, meaning we can cancel this command and make these layers permanent should we wish.

37. Click Cancel.
38. Click on the symbol to the right of 100 m buffer layer in the Layers Panel (it is called the save scratch layer). Although you may have difficulty distinguishing between the two buffer layers, which relates to the original consideration of confusion when working with temporary layers.

Data Management 95

This will open the Save Layer dialogue box, and we can go through the motions of saving our layer as a permanent file.

39. Save this layer to the GeoPackage we have created in this chapter.
40. Save the Project and close QGIS.

Now let us find out what happens when we open this map.

41. Open QGIS and under Recent Projects we should be able to double click on Ch6 and open it.

The temporary layer is still visible in the Layers Panel, but not in the project. If you open the attribute table of the temporary layer, you will see that it is empty. A stark warning to remember to save all layers we may need later. Next, we are going to explore some more data.

42. In QGIS, click the Open Project button and open the existing project in Ch6_Data_Part2 called Ch6_Part2.qgz.

A warning box appears that states there are unavailable layers, as shown in Figure 6.14. This can happen quite regularly in GIS when layers are being saved across different folders. If one layer is removed, deleted, or saved in a different sub-folder that is not transferred across to a new computer or device, then the link to that data source will break and the layer will be unavailable. This is a very common occurrence when working on GIS projects in a communal space such as a computer lab. In these instances, provided the data is not completely lost, we have some options. We must tell QGIS where this layer is now stored, which will be the case if the data layer or folder has been accidently moved. This process will fix any broken data source links.

43. Click on the layer in the dialogue box so it is selected and turns blue.

The option for Browse should appear. If we know where the layer is saved, we can navigate to it. If we do not know where the layer is saved, then we can attempt to use Auto-Find, which will search a series of folders trying to identify a file of the same name, although this can be problematic if you have multiple files of the same name.

FIGURE 6.14 Screenshot of unavailable layers in the Ch6_Part2.qgz project.

In our case, I have simply moved the shapefile to Ch6_Data_Part3, so we can browse to that location and specify the link.

44. Click on Browse.
45. Navigate to the Folder Ch6_Data_Part3.
46. Double click on the Global_Landslides.shp.

This new link should be in the Datasource.

47. Click Apply Changes.

We now have a project with two layers, soil type and global landslides. This is data from Chapter 16, where we use geoprocessing to identify landslide risk in Peru.

If we unexpectedly close the dialogue box without fixing the broken data source link, the layer will remain in the Layers Panel, but a lack of data source will be indicated by the Unavailable Layer triangle in the Layers Panel next to the layer. We can fix the broken link manually as well by clicking on this button.

If we cannot see the layer, we may have accidently specified the wrong data source. We can rectify this by right clicking on the layer in the Layers Panel > Set Data Source. We can then repeat the process of navigating to the data file within our folders.

6.4 FIXING GEOMETRIES

Another thing we can encounter a lot within GIS is geometry errors. These errors can be introduced as we progress through geoprocessing, digitizing, and editing our spatial data. They are also particularly common when we convert raster data to vector data, in part due to the intersection that can occur with the square grids of rasters. There should not be any instances in this book where a dataset has invalid geometry, as in these instances it has been resolved through pre-processing. However, as one of the aims of this book is to generate competency in QGIS, this is a data management tool that is useful to demonstrate.

SoilType is a polygon feature layer of soil type that was generated from a raster layer, which as just mentioned can be prone to geometry errors. We can check whether this layer has any.

48. In the Processing Toolbar, navigate to Vector Geometry > Check Validity.
49. Specify SoilType as the Input Layer.
50. Ensure that GEOS is ticked.
51. Keep the rest of the layers as default and create temporary outputs.
52. Ensure your dialogue box resembles Figure 6.15. Click Run.

This tool creates three new temporary layers. There is a polygon feature layer of Valid Geometry, one of Invalid Geometry, and a point file of Error Output.

The results are shown in Figure 6.16. All the polygons in blue have valid geometry, while the orange polygons have invalid geometry, with the points indicating the specific location of the error.

Data Management

FIGURE 6.15 Screenshot of parameters needed for the check validity tool.

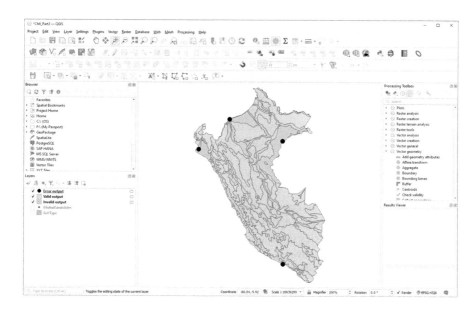

FIGURE 6.16 Screenshot of check validity for the SoilType layer.

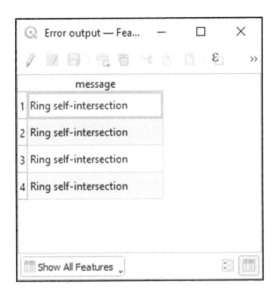

FIGURE 6.17   Screenshot of error output attribute table.

53. Open the attribute table for Error Output.

Your table should resemble Figure 6.17, and here we can see that there are four polygons that have invalid geometry, each representing a specific point location, where the polygon intersects itself.

54. Select the top row and click on the Zoom Map to the Select rows button.

This should take us to the northern coast of Peru, where we can see that the polygons have intersected themselves to maintain the gridded square shape. We may need to zoom closer to the location to really see this, shown in Figure 6.18.

These intersections cause issues with the geometry that will make geoprocessing impossible to complete. Fortunately, there is an in-built tool in QGIS that fixes these invalid geometries.

55. In the Processing Toolbox, navigate to Vector Geometry > Fix geometries.
56. Specify SoilType as the Input Layer.
57. In the fixed geometries layer, click the ... and select Save to GeoPackage. We do not currently have a GeoPackage to work from, so create a new one in your working directory called Ch6_Part2.
58. Specify the name of the layer as SoilType_Fix.
59. Ensure your dialogue box resembles Figure 6.19. Click Run.

If we repeat the steps associated with Check Validity, we should not have any invalid geometries. This layer is now ready to use within geoprocessing, and we will return to use this fixed layer in Chapter 16.

Data Management

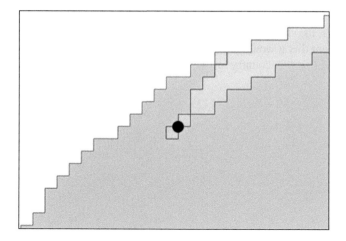

FIGURE 6.18 Visualization of the invalid geometry where the polygon has intersected itself due to the square nature of the features.

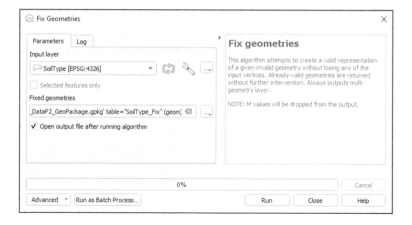

FIGURE 6.19 Screenshot of the parameters needed for the fix geometries tool.

Finally, when we undertake geoprocessing, some layers may not have a spatial index. This is simply a method to increase the efficiency of accessing and rendering spatial objects and can be easily applied.

60. In the Processing Toolbox, navigate to Vector General > Create spatial index.
61. Open the tool and specify Global Landslides as the input layer.
62. Click Run.

This will generate a spatial index for the layer, allowing any queries to be undertaken much quicker. In Chapter 16 when we return to Peru, we create a polygon from a raster representation of slope. This resultant layer will not have a spatial index. In this instance, the study area and feature layer are not overly large, meaning if we did not implement this it would not significantly slow down our work, but if we were to scale the analysis to a country level, it would allow for a substantial improvement.

6.5 CHANGING TABLE FIELDS

As alluded to in the previous chapter, in future iterations of QGIS we will be able to specify the type of field factor when we add the tables using Delimited Text Files. However, until that time, we must manually create new attributes that have the correct format, specifically related to Data Type.

63. Navigate through the tab Layer > Add Layer > Add Delimited Text File.
64. Navigate to Ch6_Data_Part_3 and add the CSO_Data.csv. This is the non-spatial layer we used in the previous chapter. Ensure 'Detect Field Types' is unticked.
65. Right click on the layer in the Layers Panel to open the Properties, and select the Fields tab.

Your dialogue box should resemble Figure 6.20. In the Type name column, we observe that QGIS reads all the fields as text format, meaning our assumption in the previous chapter was wrong. If the Detect Fields option was ticked when adding

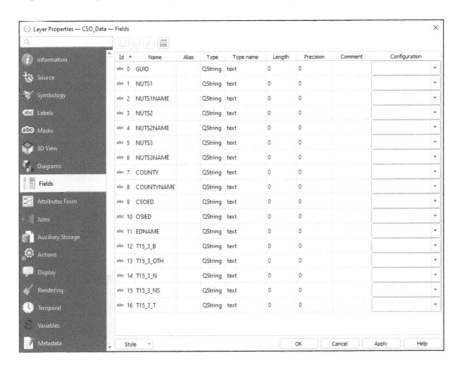

FIGURE 6.20 Screenshot of the fields tab for the CSO_Data.

Data Management   101

this layer, the Type name should be integer. For most integer attributes, using detect field is sufficient, but there are instances where it is not, and the following steps will demonstrate how to overcome this. This is important if we want to visualize this data to create maps or perform any quantitative analysis, as the attribute data is in the wrong format. We have two options here. We could use the field calculator to create a new attribute, specifying the existing field in the expression. This is what we did in the previous chapter when we created a new field based off of these values. Alternatively, as we do not need to create any new values, we could use the Refactor fields tool.

66. In the Processing Toolbox, navigate to Vector Table > Refactor fields, and open the tool.
67. Scroll down the table to T15_3_B and change the Type from Text to Integer (32 bit).
68. Repeat this for the other T15 attributes.

This tool does not update the fields in the existing layer, meaning we need to save it. For now, a temporary layer will suffice.

69. Keep the Refactored tab as [Create temporary layer].
70. Ensure your dialogue box resembles Figure 6.21. Click Run.

This creates a temporary layer called Refactored, which is fine for now as we are simply demonstrating the tool. As always, we should check this has worked.

71. Open Properties for the new layer, and click on the Fields tab.

We should see that the T15 attributes are now all Integers.

6.6 MANAGING DATA FOR SUCCESS

Managing spatial data within GIS software can be highly rewarding, as it provides a streamlined structure to enhance our GIS analysis and presentation. This chapter has highlighted some key components of data management that will be needed to maximize understanding of GIS as we complete the book. Understanding the difference between GeoPackages and shapefiles, and what this means to file management is central to GIS. The preferred usage of these formats can be hotly debated by GIS users, but in an introductory book where GeoPackages provide a streamlined format, they are preferrable. Despite that, we used shapefiles in the first few chapters to provide the learning experience, as well as the fact that certain tools in later chapters require us to work with shapefiles.

Throughout the remainder of the book, instructional steps refer to saving the resultant layer in a GeoPackage or working directory. This essentially means either the folder or GeoPackage that was download from the website. If we are working on multiple devices, such as at a university or school, it is again preferable to always use GeoPackages as this makes transferring across devices easier.

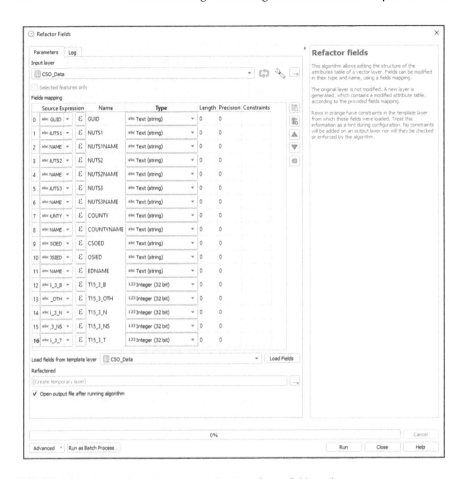

FIGURE 6.21 Screenshot of parameters for the refactor fields tool.

Temporary layers provide the flexibility to explore and investigate tools; however, they can also be problematic, especially during the experimental testing of parameters, which we undertake in subsequent chapters. Therefore, take care not to rush through the tools, generating several temporary layers. The naming conventions used here can seem overly long in places, but they are important. This is something that occurs in a lot of my courses, where the default names are used in the tools, but then when students return these layers at a later time or date, they are either not present or they cannot remember which file refers to the distance/density/ method they require. Especially when we reach Section 5, this becomes even more important.

We have explored how to navigate project files when data is missing, which will also be a key challenge that many of us will face at some point in our GIS careers. Once you are aware of this issue, it not only reinforces the need to be specific with naming conventions but becomes a lot easier to fix. Similarly, fixing vector geometries and adding spatial indices are two tools that are simple but effective in working successfully with GIS. This will be especially true when you start working independently with GIS data that has not necessarily been cleaned up or pre-processed.

This is the end of Chapter 6 and Section 2. The decision to introduce these concepts at the end of this section rather than the beginning was deliberate, as it has allowed us to get to grips with the fundamental concepts while working away in GIS to solve some real-world problems. In my experience if the concepts introduced in this chapter precede the previous three chapters, this can impair learning objectives, particularly for beginners of GIS. Having completed this section of the book, you can now consider yourself au fait with the fundamentals of GIS. We continue to build on these skills throughout the book as we begin to touch upon more advanced material. Where necessary, please do refer back to these chapters, as we have covered a lot of material, information, and tools so far.

REFERENCES

Batjes, N.H., 2016. Harmonised soil property values for broad-scale modelling (WISE30sec) with estimates of global soil carbon stocks. *Geoderma*, 2016(269), pp. 61–68. doi: 10.1016/j.geoderma.2016.01.034.

Central Statistics Office (CSO) (2018). *Census 2016 Small Area Population Statistics*. Available from: https://www.cso.ie/en/census/census2016reports/census2016smallarea populationstatistics/. Accessed March 1, 2022.

Kirschbaum, D.B., Adler, R., Hong, Y., Hill, S., and Lerner-Lam, A., 2010. A global landslide catalog for hazard applications: Method, results, and limitations. *Natural Hazards*, 52(-3), pp. 561–575. doi: 10.1007/s11069-009-9401-4.

Kirschbaum, D.B., Stanley, T., and Zhou, Y., 2015. Spatial and temporal analysis of a global landslide catalog. *Geomorphology*, 249, pp. 4–15. doi: 10.1016/j.geomorph.2015.03.016.

Section III

Cartography

7 Location and Thematic Maps

7.1 INTRODUCTION AND LEARNING OUTCOMES

In this chapter, we undertake the initial steps of cartography, primarily through the creation of location and thematic maps. GIS has supported the development of cartography through the availability of easy-to-use tools that can create professional maps. Maps enable us to facilitate a spatial understanding, allowing us to store information, reveal patterns, and illustrate relationships, providing an aesthetic and powerful representation of the questions we attempt to answer. There are several different types of maps that vary depending on the geographic features that need to be visualized.

Location or Category Maps – These are some of the simplest representations of spatial data within cartography, and subsequently the most widely used. To represent location, each feature is represented in the map, either as a point, line, polygon, or raster. To facilitate spatial understanding, data is represented using specific attributes. For categorical fields, each value is assigned its own unique color, which results in a map of (hopefully) distinctly different colors. This type of visualization is very effective when we have categorical data as it supports differentiation among attributes and facilitates understanding of where features can be found. Simple examples include symbolizing lines representing roads based on whether they are motorways, primary roads, or secondary roads, or polygon features representing soil type, with each unique soil visualized as a separate color.

Thematic Maps – These maps represent the spatial pattern of a geographic feature within a specified area. Thematic can be simply defined as being related to a subject matter, meaning the geographic feature or subject matter of the study is the theme of the map. There are several different types of thematic maps, including graduated symbols, choropleth, dot distributions, change, and flow maps. In this section, we develop graduated symbol maps in this chapter, choropleth maps in Chapter 8, and change maps in Chapter 9.

Graduated Symbols – In these maps, shapes of different sizes are used to simplify and represent more complex patterns, such as amount. In this chapter, we represent the amount of litter found at specific locations, with points containing more litter represented by larger points. This provides a visual aid to support spatial cognition, as patterns between quantities and location can be discerned almost immediately. Such maps have also been referred to as proportional point symbols and bubble maps.

DOI: 10.1201/9781003220510-10

Choropleth Maps – These maps use a graduated color palette (or symbology) to represent quantitative data that are aggregated over predefined regions, such as countries, states, or counties. In Chapter 8, we use cartography to investigate global school closures, representing countries which had a larger disruption of education during the Covid-19 pandemic visualized using darker colors. Color palettes are often distinguished by hue or lightness to support spatial cognition of which locations have larger or smaller aggregated values.

Change Maps – These maps are a subset of choropleth maps, as these again occur over an aggregated geographic unit. However, the main feature that is visualized pivots around a central value, such as zero. Examples include land cover change, with a percentage increase or decrease over time being visualized. We develop change maps for forest cover in Chapter 9. Again, color palettes are distinguished by hue or lightness, but the palette itself is divergent as opposed to sequential.

Heat Map – These maps represent the density of geographic features contained within our study area. Areas of higher density are represented as clusters, and subsequently visualized as 'warmer' in the symbology, while areas of lower density are visualized as 'colder'. This is particularly useful when our spatial data is voluminous and dense, and discerning an immediate pattern based on location, category, or attributes is difficult. Therefore, such a mapping convention describes the spatial pattern of the data more than previous methods where the primary visualization is on both the spatial and attribute data. We create heatmaps in Chapter 13 as part of our work on density analysis.

In this chapter, we consider SDG14, specifically target SDG14.1, which aims to prevent and significantly reduce marine pollution. Therefore, this chapter details how to visualize macro-plastic pollution, and then how to effectively present such work. To achieve this, we use data from a citizen science project called OpenLitterMap (https://openlittermap.com/world) that aims to map and monitor global litter (Lynch 2018). OpenLitterMap data is published in the public domain but remains copyright to OpenLitterMap and Contributors. Such monitoring programs are vital in supporting sustainability, as the information generated can support mitigation strategies such as targeted clean-ups, placement of refuse bins, and education programs. Here we create location and thematic maps to maximize this information and inform decision making to support SDG14.1.

OpenLitterMap is a citizen science project, which means that members of the public can upload data, photos, and spatial locations of litter. Because of this, there are inevitably gaps in the data collected, meaning what we present is an incomplete picture and is by no means representative of a systematic survey. Due to the sparsity of data, this chapter focuses on Puerto Rico, where there is relatively good spatial coverage across the east of the island. Regardless, we should think critically when interpreting the map, especially as there is no spatial structure to data collection; however, such methods of data collection permit global efforts of collecting information on features such as plastic, the likes of which have not been historically possible.

Location and Thematic Maps   109

To facilitate the map-making process, the data has been downloaded and pre-processed. The 'type' of litter has been aggregated into categories and quantified as the 'total' amount of litter identified at a location. Should you wish to explore the data for other locations, navigate to the website, scroll to a specific country, click Download, and enter your email. In about 5–10 minutes, you receive an email with the link to the data. However, to follow this chapter on a different location, the pre-processing needs to be completed. Subsequently, for the remainder of the chapter, please work off the *.csv provided. By the end of this chapter, you will have completed three learning outcomes:

- Develop multiple presentations of litter data using different symbologies
- Appreciate the importance of the key cartographic elements, including map balance, scale bars, north arrows, and legend
- Use the Print Composer to create a professional map output

7.2 CASE STUDY: SDG14.1 MAPPING PLASTIC POLLUTION IN COASTAL SETTINGS

Firstly, we need to import the *.csv into QGIS.

1. Open a New Empty Project in QGIS.
2. Navigate through the tab Layer > Add Layer > Add Delimited Text Layer.
3. Add the Ch7Data.csv file.

The main difference with this dataset compared to previous chapters is that this data already has its geometry (or coordinates) included. When adding the data, we can specify this.

4. In the Add Layer dialogue box, navigate to Geometry Definition, select Point Coordinates.

The X and Y fields should automatically populate with lon and lat, respectively. If they have not, manually complete this. Lon stands for longitude, and lat stands for latitude. Refer to Chapter 4 for a refresher if needed.

5. We need to define the projection. The data have been collected using a GCS of WGS 84. Therefore, select EPSG:4326 – WGS 84.
6. Ensure Detect Field Types is ticked and your dialogue box resembles Figure 7.1.
7. Click Add and close the dialogue box.

The data should be added to the project and resemble Figure 7.2. As this is a citizen science project, it is continually being updated. Therefore, over the course of time, more locations will be added. This data was downloaded in November 2021, so if working with more recent data, there may be a different pattern. We should be able to make out the rough shape of the USA, in particular the eastern coast, such as the

FIGURE 7.1 Screenshot of the parameters for add delimited text file with geometry definition.

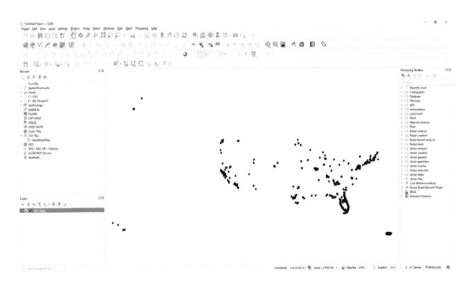

FIGURE 7.2 Screenshot of the QGIS map canvas with Chapter 7 data loaded.

Location and Thematic Maps

Florida peninsula. At this point, it would be useful to add a basemap for our own reference.

A basemap is itself a type of map that allows us to reference our thematic information. It contains a detailed representation of the physical and/or anthropogenic features on the ground, with various designs each styled for a specific purpose. Examples can include ocean topography, satellite images, or road networks. For our purposes, we are interested in simply aligning ourselves with the geography of the region, as well as certain towns and locations. Therefore, a basemap that best represents the natural features such as land and water, as well as settlements and roads is important. The OSM layer we have used thus far in this book does exactly this.

8. In the Browser Panel, navigate to XYZ Tiles and select OpenStreetMap. If necessary, reorder the layers in the Layers Panel so that the points are visible.

Your screen should resemble Figure 7.3. Technically, we have now created a simple point location map of reported litter in the USA, meaning we could theoretically stop here. However, cartography is all about best representing the spatial pattern and uncovering hidden or unknown processes. It would also be quite a short chapter if we were to stop at this point, and we would not be best supporting efforts to reduce marine pollution through cartography. Moreover, in this map there is a lot of noise, particularly related to missing data and the basemap. If we were to complete the presentation of this data at this stage, we need to think about how best to represent the primary and secondary information principles.

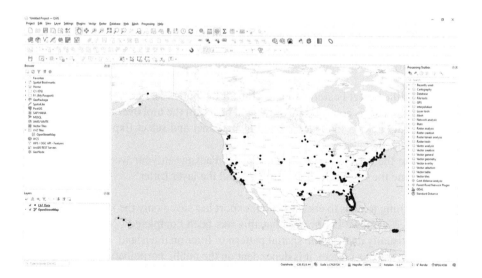

FIGURE 7.3 Screenshot of the QGIS map canvas with Chapter 7 data and OSM Basemap. Basemap is the OpenStreetMap XYZ tiles which is © OpenStreetMap contributors and available under the Open Database License. Please see https://www.openstreetmap.org/copyright.

The primary information is what we want to show, in this case the distribution of litter collected in the USA. The secondary information provides reference for our primary data but is not the focus of the map. The basemap included contains a lot of features, including roads, towns, and labels that our map does not require. Therefore, we may want to consider an alternative political boundary as our base layer for presentation purposes, such as US administrative boundaries. Moreover, while we have spatially interrogated the data, we have not yet explored the attributes associated with the data in detail. This is important as the attribute table may contain additional data that we can use to best present the information available to us.

9. Before we progress, save the project as Chapter 7.
10. Open the attribute table.

There are several attributes, including the primary type of litter that was found at the location, as well as the total amount of litter items across categories, which include smoking, food, coffee, alcohol, soft drinks, sanitary, dumping, industrial, coastal (which includes fishing gear), and other. By mapping the type of litter and the amount found at the location, mitigation efforts can be more targeted. For example, an increase in refuse bins in an area that is dominated by fishing gear would not have the desired effect as the litter is washing in from the sea, but a specialized cigarette butt waste bin in an area dominated by smoking litter would be more effective. To maximize the impact of our work using the attribute data, we now focus on Puerto Rico where there is a high amount of data in a relatively localized geographic area.

11. Using the zoom tool, zoom to Puerto Rico.

Use your geography knowledge and the basemap to complete this (hint: it is in the south-east of the map, in the Caribbean). To make our analysis simpler, we want to extract only the data associated with this island.

12. Use the Select Features tool to grab all the points on Puerto Rico and select them. This is achieved by drawing a box with the tool over the island.
13. Right click on the layer in the Layers Panel and select Export Data > Save Selected Features As.
14. Save the selected points as a new GeoPackage in your working directory. Name the GeoPackage PuertoRico and the layer Litter.

The first thing to do is to check that our export has saved only the locations on Puerto Rico. Once we have confirmed that this has been completed correctly, we should begin to visually analyze the spatial pattern of litter on the island. Remember, as this is a citizen science project, it is dependent on the public participating. Therefore, it is far from a complete representation of litter, as well as a static representation in time. The most obvious interpretation of this data is that there is a clear east-west divide to the island in terms of data collected. This does not mean there is less litter in the west, but simply that we have more information in the east. This allows us to make

Location and Thematic Maps

more informed decisions on where to perform targeted clean-ups based on more heavily sampled areas but does not mean efforts should not be focused in data-poor areas.

When users submit records to this project, the litter is categorized by type. During pre-processing, a new attribute called PRIMARY was created, which refers to the predominant litter type found at each point location. Furthermore, several integer fields for all types of litter were created to provide quantitative information on the amount of litter. We know from Chapter 6 that our fields may not be detected as integers, so we should first check this.

15. Open Properties for Litter and navigate to the Fields tab.

If all the fields after 'total_litter' are recorded as Integer, with the exception of PRIMARY, the data is correct. If they are text, then we need to refactor the fields.

16. If the attributes are text, navigate in the Processing Toolbox to Vector Table > Refactor fields, and change all attributes after 'total_litter' (including it) to integer, with the exception of PRIMARY that should remain as text. Save the resultant layer in the GeoPackage.
17. Confirm this has worked by navigating back to Fields in Properties.

Now would be a good starting point to visualize the primary type of litter identified at each location on the map. We do this by changing the symbology.

18. Right click on the Litter layer, and select Properties.
19. Select the Symbology tab.
20. Change Single Symbol to Categorized.
21. Select PRIMARY as the value.
22. Select a qualitative color ramp. For now, we can stick with the default of random colors.
23. Finally click Classify, and we should see the different categorizations appear.
24. Ensure your dialogue box resembles Figure 7.4 (note you may have slightly different colors, as they are random). Click Apply, and then close the dialogue box.

The data is now symbolized based on the primary litter type. You should have a map that resembles that of Figure 7.5.

25. In the Layers Panel, expand the list of values within Litter, which for now can act as a type of legend.
26. Try unticking some of the layers and types of litter in the Layers Panel. When these are unticked, they disappear in the map.

We have now created a more detailed location map, representing the primary type of litter found in each location. Again, we could stop here, but cartography is the art

FIGURE 7.4 Screenshot of parameters to fit a qualitative symbology for PRIMARY litter type.

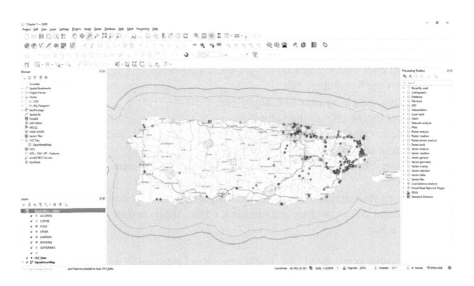

FIGURE 7.5 Screenshot of random color symbology for PRIMARY litter type in Puerto Rico. Basemap is the OpenStreetMap XYZ tiles which is © OpenStreetMap contributors and available under the Open Database License. Please see https://www.openstreetmap.org/copyright.

Location and Thematic Maps

of uncovering hidden information to support the end goal, in this instance to support mitigation strategies to reduce marine pollution. Therefore, the amount of litter could be a more useful attribute to map, as strategies could be targeted to maximize the amount of litter removed from the location.

Using the attribute table, we can see that the amount of litter (total_litter) recorded at each location ranges from 1 to 157. We cannot visualize this value by simply changing the symbol, instead we must create a thematic map. Thematic maps use information from the attribute table, in this case total amount of litter, to create a symbology for a layer.

27. Right click on the Litter layer, and select Properties.
28. Select the Symbology tab.
29. Change Categorized to Graduated.
30. Select total_litter as the value to symbolize.
31. Change the Method from Color to Size.
32. Change the Classes to 10.
33. Change the method from Equal Count to Natural Breaks.
34. Click Classify.
35. Change the Size to minimum 2 and maximum 10.
36. Uncheck Link Class Boundaries.
37. Ensure your dialogue box resembles Figure 7.6. Click Apply and then close the dialogue box.

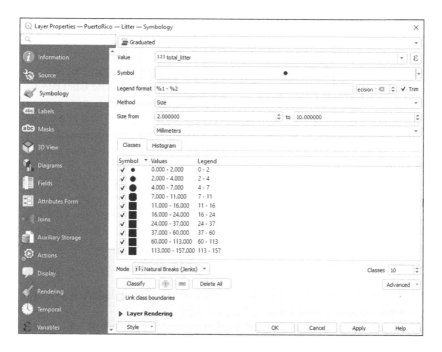

FIGURE 7.6 Screenshot of parameters to symbolize total_litter using graduated symbols.

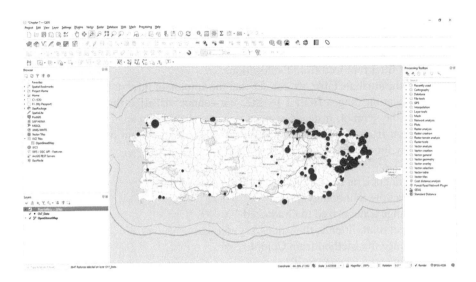

FIGURE 7.7 Screenshot of the initial graduated symbology based on total_litter. Basemap is the OpenStreetMap XYZ tiles which is © OpenStreetMap contributors and available under the Open Database License. Please see https://www.openstreetmap.org/copyright.

Sometimes the 'size from' values will automatically update when we change another section in this dialogue box, hence why the instructions were ordered to change the size last. If you do not have variation in your output as in Figure 7.7, double check this.

Before we progress, a quick note on the total amount of litter in Puerto Rico compared to that of the complete dataset of the USA. The maximum value of individual litter pieces in both datasets is 157, meaning the maximum value in the symbology would have been the same whether we had used the Ch7Data layer or the Litter layer. However, in an instance where the maximum litter or types of litter in the study area do not match the full range of values in the complete dataset, we would have an incorrect symbology that is visualizing values beyond our range. When undertaking such steps to generate maps and present data, we must keep this mind.

We now have more information provided in our map shown in Figure 7.7, but it's still quite difficult to read. To make it easier to identify large litter deposits across the whole island, it would be a good idea to remove the smaller deposits. To do this, we could use the Select Features by Values button, but this will only select the features. Instead, we want to filter this layer, such that those features that we are not interested are only temporarily removed.

38. Navigate through the tab Layers > Filter (note, we can also navigate here by right clicking on the layer in the Layers Panel).
39. Specify any locations that are equal or greater than ten pieces of litter associated with each point using the following expression: "total_litter" >= 10.
40. Ensure your dialogue box resembles Figure 7.8. Click OK.

Location and Thematic Maps

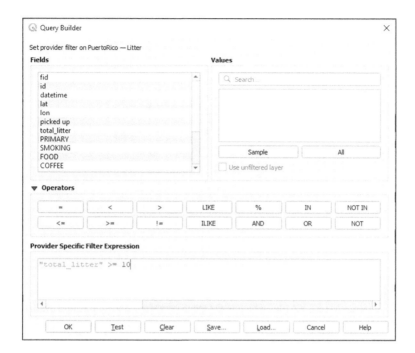

FIGURE 7.8 Screenshot of the parameters for the filter tool.

This filter specifies that only points with ten or more pieces of litter are returned.

41. Zoom to Húcares (on the east of the island, near the Reserva Natural de Humacao). This location is shown in Figure 7.9 for reference.
42. Turn on the unselected layer of all points from the USA.

When comparing the two layers, one which is symbolized using graduated points and has a filter applied, with the other that is not symbolized at all, we can begin to see the difference in location of smaller deposits of litter and the larger ones.

43. Turn off the layer of all points from the USA.

There are still two issues with this map. The first is that the point features are stacked in the visualization and the second is that some classes appear to contain multiple values (i.e., 0–2 and 2–4). For example, it is unclear how a point with two pieces of litter would be symbolized. It technically could be in both the 0–2 and 2–4 categories. Unfortunately, there is no clear method to overcome this other than to manually change the classification values ourselves.

44. Reopen the Properties dialogue box and navigate to the symbology tab.
45. Double Click on the 2.000–4.000 under the values in the classification tab.
46. This should open another dialogue box. Change 2.000 to 2.001.

FIGURE 7.9 Screenshot of the graduated symbology near Húcares, Puerto Rico. Basemap is the OpenStreetMap XYZ tiles which is © OpenStreetMap contributors and available under the Open Database License. Please see https://www.openstreetmap.org/copyright.

FIGURE 7.10 Screenshot of the enter class bounds function.

47. Ensure your figure resembles Figure 7.10. Click OK.
48. Repeat this for all the ten classes. This way, there will be no ambiguity as to where the value of 2 will be visualized. The complete list of value intervals is shown in Figure 7.11.

The issue of stacking stems from the fact that all features are rendered/drawn at the same time. This means they overlap each other, meaning that there may be points that are not observable because of the render order. Stacking is a common default in GIS but can be problematic as features are hidden and subsequently missed from interpretation. We overcome this by providing an order for which the categories are symbolized.

49. Click on the Advanced button in the Symbologies tab, and select Symbol levels.

Location and Thematic Maps

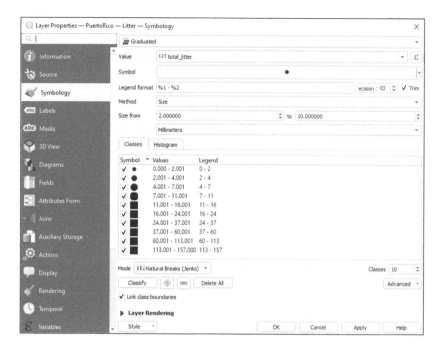

FIGURE 7.11 Screenshot of the manually changed class boundaries.

50. Tick Enable symbol levels, which should activate the table.
51. In the second column, choose the order in which the layers are rendered, with the larger symbols rendered first. Start in the first row (i.e., the smallest symbol) and specify 10, then 9, then 8, and so on, until the table resembles Figure 7.12.
52. Click OK then Apply.

There are now no symbols representing the larger values that are overlaying and covering any of the smaller deposits.

For officials, volunteers, and organizations, such a map might support local clean-ups. While every piece of litter can have a significant negative impact on the landscape, targeting locations with more litter will support SDG14, specifically SDG14.1 to reduce marine pollution, particularly in a coastal area such as this. While a useful map, we could further uncover hidden patterns by combining the type and amount of litter. In the attribute table, for each litter category there is also a number reflecting the amount of that type of litter.

53. Open the Filter option to clear the current selection.

Another problem with the previous classification (well done if you spotted it) was that values of zero were being represented in the symbology. This was not really an issue in the previous visualization as we filtered out all values below ten. However, moving forward we need to rectify this. The next map will visualize alcohol waste, meaning

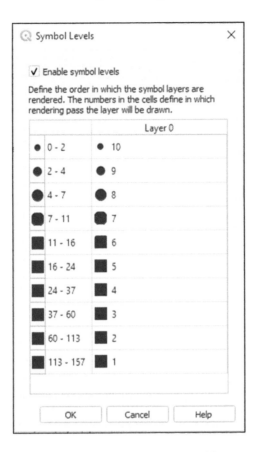

FIGURE 7.12 Screenshot of the symbol levels to prevent stacking.

we must filter observations of zero, as this would be somewhat counterintuitive to anyone reading the map.

54. Open the Filter option and specify the following expression: "ALCOHOL > 0".

Once this filter has been applied, we want to symbolize the remaining data points.

55. Repeat the steps to symbolize total litter, but this time use ALCOHOL as the value with nine classes. However, to make the map more aesthetic, we may want to change the color of the symbols to red. Remember to reorder the rendering in the advanced options and manually change the classifications to include the 0.001 difference as these features will reset. Your dialogue box should resemble that of Figure 7.13.

Finally, we have discussed the 'noise' associated with the choice of basemap earlier in this chapter. An outline of Puerto Rico or an alternative basemap may provide less noise and act more effectively as a second-order principle for reference. A satellite

Location and Thematic Maps

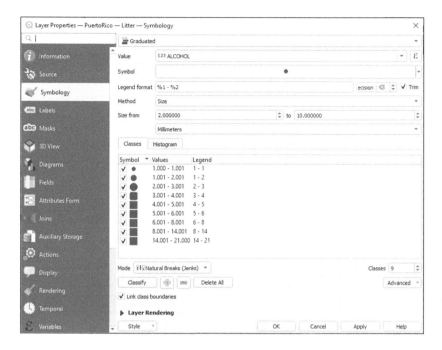

FIGURE 7.13 Screenshot of the parameters to symbolize ALCOHOL litter.

image should provide the land and sea delineation, as well as highlight any built-up areas that we may need to consider in interpretation.

56. Click on the QuickMapServices Button that we installed in Chapter 2 as a plugin.

If this has not been completed, navigate through the tab to Plugin > Manage and Install Plugin and install the QuickMapServices plugin.

57. Select Search QMS.

This should open a docked dialogue box in the bottom right-hand corner.

58. In the search bar, type 'satellite'.

This should return all basemap layers that contain satellite imagery as shown in Figure 7.14. Because these are not part of the core QGIS functionality, some of these layers may disappear in the future due to a lack of continued maintenance. Therefore, we can be flexible in terms of which one we choose. In this instance, choose Google Satellite.

59. Navigate to Google Satellite and click Add.
60. Turn off the OSM basemap in the Layers Panel.

FIGURE 7.14   Screenshot of the search QMS panel.

To create a professional map output, we use the Print Composer. The Print Composer opens a separate area in QGIS, which is linked to our initial project and data. This allows us to design a map, while also managing and editing the spatial data, if we so desire.

61. Click on the New Print Layout button.
62. A dialogue box will open requesting a name for this map. Call it 'Litter Map'.
63. Click OK.

A new window that resembles Figure 7.15 should appear.

64. Click on Add Map on the vertical toolbar in the Layout.
65. A black cross should appear. Draw an outline of where the map will go. Aim for approximately 75% of the area starting from the left-hand side.

A map appears that is linked to the map canvas. We can change the size and shape of this by moving the small white squares in the corners if needed. It is important to note here that layers turned on in the map canvas appear in our layout. Therefore, we only want layers turned on that we want in our final map.

Location and Thematic Maps

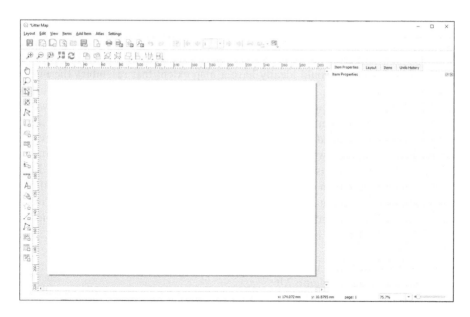

FIGURE 7.15 Screenshot of the print composer.

FIGURE 7.16 Example of map balance, with the same data presented at three different spatial extents. Basemap is Google Satellite which is Map data ©2015 Google, please see https://www.google.at/permissions/geoguidelines/attr-guide.html.

66. Remove all layers except for the Litter layer and Google Satellite basemap by right clicking on them and selecting remove layer in the map canvas.

Something to consider at this point is map balance. Map balance is where we arrange the details of the map to maximize their effectiveness. In cartography, empty space is where there is an abundance of emptiness on our map, often termed white- or blue space because of the default white background in many GIS programs or the blue of the oceans in basemaps.

If we take the example in Figure 7.16, the first image is zoomed out such that we are observing data not only for Húcares but also most of the eastern coast of the island. In the second image, we are observing data for the town and the surrounding nature reserve, albeit with a lot of blue space due to the coastal nature of the dataset. In the final image, we are zoomed in to the town, specifically a single street. Depending on the purpose of the map, we might choose different extents.

For a map output that aims to cover the town and surrounding area, the second image would be preferred, although we must note that the presence of blue space could be problematic, and we revisit this shortly. Therefore, to maximize the balance of our map, we would ideally have a map extent that resembles the middle image in Figure 7.16.

When we draw the outline of the map, it renders at the extent at which we are viewing the data in the map canvas. Therefore, prior to drawing the outline, ensure that the extent is specified as such. We can manually add the coordinates to the extent in the Print Composer, which we revisit in the next chapter. Manually changing the extent in the map canvas using a combination of Zoom In, Zoom Out, and Move Item Content buttons can be somewhat trial and error, but it can provide flexibility in highlighting specific features. Alternatively, we could select the Move Item Content button in the Layout Vertical Toolbar selection as this allows us to zoom and pan around the map directly in the Print Composer.

67. Specify the extent using one of the above-mentioned methods to match the second image in Figure 7.16.

At this stage in our GIS careers, we want to be able to create a map that visualizes our data while adhering to standard cartographic practices. There are certain elements that each map should have. Therefore, we want to create a map that has (1) a scale bar, (2) a north arrow, and (3) a legend. A scale bar provides a measurement aid to readers of the map navigating the landscape and data. In essence, a real-world distance e.g., 1 km, is represented using a map-distance, e.g., 1 cm. A north arrow simply allows readers to orientate themselves in a specific direction. A legend is necessary as it provides a key to be able to interpret the map, with symbols or colors labeled to aid readership.

One observation from having taught GIS over several years is not to make the north arrow and scale bars prominent features of the map. Where map outputs have a north arrow or scale bar as large as the map itself, it shifts the focus from the geographic features and patterns to the orientation or measurement of the map. Therefore, when placing these features take care to ensure they are visible and legible, but do not shift the balance of the map. In many respects, we need to make these map features as insignificant as possible, while ensuring readability. This is particularly pertinent for the scale bar and north arrow, while the opposite is often the case for the legend, with many a legend being too small to read. On the vertical toolbar in the Layout frame, we should see the three options for adding these features.

68. Click on Add Scale Bar.
69. Again, a black cross should appear. Draw the scale bar to create its extent.
70. Click on Item Properties which should have appeared on the right of the screen. If it has not, right click on the scale bar, and select item properties.
71. Change the scale bar units to km.
72. Change the fixed width to 0.5 units.
73. Change the height from 3 to 1 mm.

The following item properties are displayed in Figure 7.17.

Location and Thematic Maps

FIGURE 7.17 Screenshot of the parameters to specify the item properties of the scale bar.

74. Add a North Arrow.
75. Click on item properties. There are a range of symbols that could be used here. Select a style of your choosing.
76. Finally, we want to add the legend. Draw a box using the tool in the remaining 25% of the space on the right-hand side of the layout, which should produce a legend of the thematic map.

We should notice that Google Satellite is present in the legend. This is not informative to the map output and acknowledgment of this layer can be included in any figure caption, meaning we can remove it from the legend.

77. Untick 'Auto Updates' in the Legend Items banner of Item Properties.

78. Right click on Google Satellite and select Hidden.

This will remove any layer that we do not want in the legend. Therefore, if other layers have not been removed, we can repeat this process so that only Litter is visible.

We may also want to remove the layer name from the legend, particularly if it is not informative to the map content. In our case, it is not, or could in fact cause more confusion. We can delete or modify this.

79. In Item Properties, ensure Auto Updates is still unticked and double click on the Litter layer. This should open a new tab.
80. There is a text box for this label. Reword this to 'Amount of Alcohol Litter' to reflect the type of litter that is present.
81. Click on the triangle to return to Item Properties.

Finally, we still have the ambiguity associated with our legend. For example, it is unclear from the legend which symbology is used for 1 (despite this being manually classified).

82. Click on the downward triangle of Litter in the Legend Items to visualize the legend values.

Double click on the 1-1, which should open a new dialogue box. Change this to 1.

As the class intervals mainly cover a single value, we can use the following legend shown in Figure 7.18. Remember, legends are designed to aid readability of our maps.

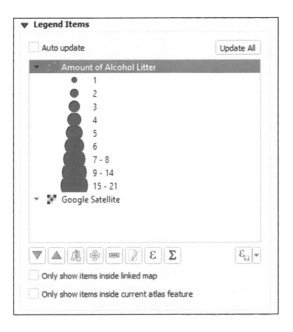

FIGURE 7.18 Screenshot of the parameters to rename the legend values.

Cartography is not an exact science, and while there are a set of standard features that are required to make it readable, everyone will have their own style regarding color, size, use of space, positioning, and balance. Therefore, take your time to make the map as aesthetically pleasing as you wish. The same way there are multiple ways to write an essay, the same can be said of maps. Therefore, the map generated by following the instructions is provided for context in Figure 7.19, but if your map differs slightly, do not be concerned.

Finally, it was alluded to earlier, but the shape of the bay has presented us with a slight problem in presenting our work. To effectively capture all the data along the coast, town, and nature reserve, we needed to be zoomed out; however, we have subsequently lost a lot of information within the town itself. We can turn this empty blue space into an advantage by adding an inset map that focuses on the main street of the town. Before we do this, however, if we edit, zoom, or add more layers to our map canvas it will update in the Print Composer, rendering all our previous steps in the Print Composer moot. Therefore, we need to lock the visualization in place.

83. Select the outline of the map in the Print Composer and open Item Properties.
84. Under Layers, ensure that Lock Layers and Lock Styles for Layers are ticked, as shown in Figure 7.20.
85. Return to the map canvas and zoom into the main strip of Húcares.
86. In Print Composer, click on Add Map and draw a box in the empty blue space south of the strip.

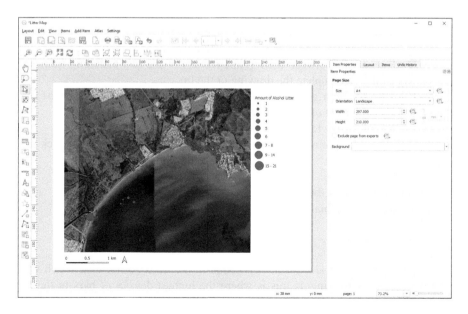

FIGURE 7.19 Screenshot of the map layout after initial positioning, with cartographic standards applied. Basemap is Google Satellite which is Map data ©2015 Google, please see https://www.google.at/permissions/geoguidelines/attr-guide.html.

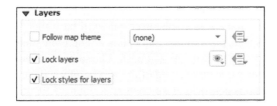

FIGURE 7.20 Screenshot of the parameters to lock the map features in place.

87. Click on the first map (Map 1) that we added. You will need to select the Select/Move Item button from the vertical toolbar.
88. In the Item Properties, scroll to Overviews.
89. Click on the Green + to add an overview.
90. Set the Map Frame to Map 2 (this is the one we have just added of the town).

The area that is represented in Map 2 should now be highlighted in the first map.

91. Change the symbology to a transparent fill and the outline from no pen to a white dash line. It would also be worthwhile increasing the stroke width to 1.

While the outline is obvious to us, we may want to connect the two maps.

92. Click on Map 2 and in Item Properties scroll down to Frame.
93. Ensure that it is ticked, change the color to white, and change the thickness to 1 mm.
94. In the vertical Layout toolbar, select Add Arrow.
95. Draw an arrow between the top left corners of the two matching frames. Right click to complete the drawing of the line.
96. Click on the item properties for the arrows (called Polyline in the list of Items on the right of the screen).
97. Change the arrow to none for the head.
98. Finally, under Main Properties, change the color of the line from black to white.

The item properties for the arrow are shown in Figure 7.21.
You should have a final map that resembles Figure 7.22.

7.3 CASE STUDY: CONCLUDING REMARKS

The highest density of alcohol litter is found in the town of Húcares, which while not necessarily surprising as this is where the highest density of bars in the area is, does quantitatively highlight the most likely source of the pollution. From the inset map, the predominant quantity of alcohol waste is individual pieces of litter, notably cans, bottles, or cups. To address SDG14.1, targeting those bars which have largest expanse of litter surrounding them with strategies such as a 'bottle/can deposit program' might encourage recycling and reuse, or incentives for the removal

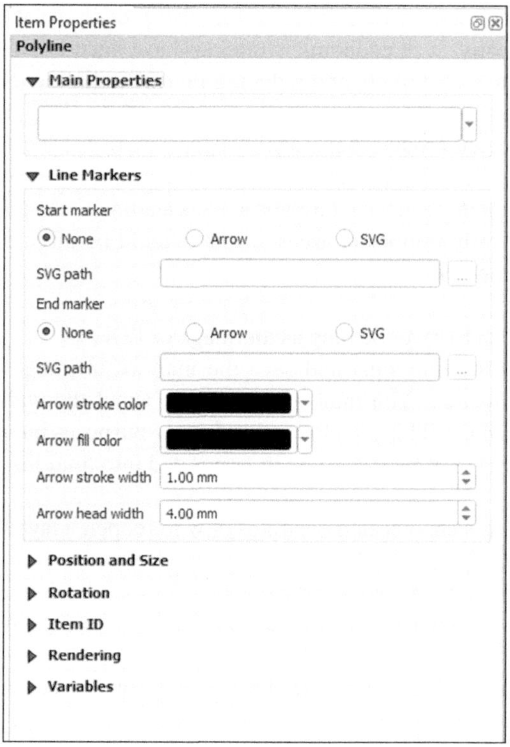

Location and Thematic Maps

FIGURE 7.21 Screenshot of the parameters to change the properties of the arrow.

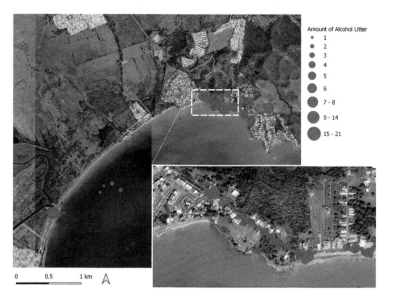

FIGURE 7.22 Final thematic map demonstrating the amount of alcohol litter in Húcares, Puerto Rico, as collected from OpenLitterMap. Basemap is Google Satellite which is Map data ©2015 Google, please see https://www.google.at/permissions/geoguidelines/attr-guide.html.

of single-use containers in these bars would also reduce the volume that could be littered. While costly, such economic strategies have been found to be successful in recycling single-use containers and reducing marine pollution (Schuyler et al. 2018; Tudor & Williams 2021). There are also isolated locations along the coastal road with litter points nucleated around them, again suggestive of bar locations. Perhaps more unexpectedly are the isolated deposits in the reserve and off the coast. These sites of litter may represent more unstructured processes leading to the pollution, but as these represent ecologically important areas, efforts should be directed to exploring why litter is being dumped here.

99. We may wish to save the map as an image or PDF.
100. Go through the motions, and save the map as an image or PDF in your working directory. You may get a warning about the width and height of the WMS image. This should be fine for our purposes but note that certain basemaps may not be able to be incorporated into final images.

The final part of this chapter demonstrates how to reopen a layout should it close.

101. Save the Litter Map layout and close it.
102. Click on the Show Layout Manager button in the Project Toolbar.

This should open a dialogue box with the saved layouts, as shown in Figure 7.23.

103. Double click on Litter Map, and it should open.

We have now successfully represented discrete locations of plastic pollution in a coastal setting, creating location maps detailing the type of litter, before using graduated symbols to generate a thematic map to illustrate the amount of litter. We have also used the Print Composer for the first time in this book, highlighting some of

FIGURE 7.23 Screenshot of the Layout Manager dialogue box.

the tools that can be used to generate professional maps. Finally, we have introduced key cartographic elements, including scale bars, north arrows, legends, and map balance. In the next chapter, we develop more thematic maps, when the data is best represented using aggregated units (i.e., country), as well as exploring different color palettes.

7.3.1 Test Yourself

If you want to test yourself on the learning outcomes of this chapter, complete the following:

a. Generate a map layout using graduated symbols for a different category of litter in Húcares.
b. Generate a map layout using graduated or categorical symbols for litter types elsewhere on the island.

REFERENCES

Lynch, S., 2018. OpenLitterMap. com–open data on plastic pollution with blockchain rewards (littercoin). *Open Geospatial Data, Software and Standards*, 3(1), pp. 1–10.

Schuyler, Q., Hardesty, B.D., Lawson, T.J., Opie, K. and Wilcox, C., 2018. Economic incentives reduce plastic inputs to the ocean. *Marine Policy*, 96, pp. 250–255.

Tudor, D.T. and Williams, A.T., 2021. The effectiveness of legislative and voluntary strategies to prevent ocean plastic pollution: Lessons from the UK and South Pacific. *Marine Pollution Bulletin*, 172, p. 112778.

8 Choropleth Maps

8.1 INTRODUCTION AND LEARNING OUTCOMES

In this chapter, we develop another type of thematic map, specifically a choropleth map. Here we address SDG4 Quality Education, focusing on how maps can be used to investigate school closures during the Covid-19 pandemic. Because of the importance of education, there has been a large amount of data collected globally on schooling since the onset of the pandemic. UNESCO (2022) has collected data that monitors the national impact of the pandemic on education systems, including the duration of school closures and enrollment figures. The Global School Closures data can be downloaded at the Humanitarian Data Exchange (https://data.humdata.org/) under the Creative Common Attribution International License. Given the implications of long school closures on the educational attainment and wellbeing of children, as well as the increased possibility of non-completion rates increasing as students learn remotely, identifying countries that may need to further prioritize education in the short- to medium term is important. As such, this chapter explores the impact of Covid-19 on global school closures through cartographic representation.

By the end of this chapter, you will have completed three learning outcomes:

- Explore the different classification methods for symbolizing quantitative attributes
- Select an appropriate graduated symbology to represent school closures
- Continue to use the Print Composer to create professional map outputs

8.2 CASE STUDY: SDG4 VISUALIZING PANDEMIC SCHOOL CLOSURES

Firstly, let us open an existing QGIS project. This is the first chapter where we are using an already existing project.

1. Open QGIS, but this time instead of opening a New Empty Project, click on the Open Project button in the Project Toolbar.
2. Navigate to the extracted folder for Chapter 8, and add the Ch8.qgz file.

This loads a pre-existing project with two layers that are represented in Figure 8.1, a spatial representation of all countries at a global scale called world-administrative-boundaries from the World Food Programme (2019) licensed under an Open Government License v3.0 and a table called Duration. Duration is the UNESCO (2022) data 'duration of school closures' downloaded in March 2022. This table has been pre-processed and changed from the original data source to ensure that the unique country code matches that of the world-administrative-boundaries. The

FIGURE 8.1 Screenshot of the Chapter 8 project.

decision to match the ISO3 codes from the Duration table to the world-administrative-boundaries dataset (and not vice versa) was taken simply due to the ease of editing the table of a *.csv file over that of a shapefile. The Duration table has also been reduced so that only attributes that are calculated in weeks are present. Upon completion of this chapter, you could return to the data portal and use the other datasets that are available, but note that you must pre-process the data to ensure a unique country code. Moreover, if you use any of the other files (i.e., enrollment) prior to importing it into QGIS you must either remove all the commas such that the numbers are recognized as integers or create new integer attributes from the existing ones that will be specified as text due to the commas.

Firstly, we need to join the table to the spatial layer. As we know from Chapter 5, to complete a table join, we must have an attribute that is unique across both layers. It is good practice to try and identify this from the attribute tables. I will provide the answer below in a moment, but it is a useful exercise to interrogate the data yourself.

3. Open the attribute tables for world-administrative-boundaries and Duration and explore them to see if there is a common attribute.

The answer is usually found in the table that we are looking to join. This oftentimes holds only the key information, especially if it has been pre-processed (like it has in this occasion). In this table there are only six columns, with four of these representing data on school closures. Therefore, the only two options are Country and ISO3. In GIS, we often use a code to uniquely represent each location, as this avoids spelling differences or representing locations across languages where subtle differences might mean a non-exact match. Therefore, while we 'could' use Country, it would be poor practice to do so, and in fact, there are spelling differences in these datasets, meaning our join would be incomplete. Therefore, based on the Duration table, the attribute we should aim to join is 'ISO3'. Next, check in the other attribute table to see whether such an attribute exists. We should note here that the attribute names do not always match (although in this case they do). Now that we have explored the data to identify the unique identifier, we join the layers.

4. Right click on the world-administrative-boundaries and open Properties.
5. Click on the Join tab.
6. Click on the Green Plus.
7. Set the Join layer as Duration.
8. Set the Join field as ISO3 and the target field as iso3.
9. Click OK.
10. Click OK on the Properties dialogue box too and close it.

We should be well used to this by now, but we should check to see if the join has completed properly.

11. Open the attribute table of world-administrative-boundaries and check that the join has worked.

Your attribute table should resemble that of Figure 8.2. There will be a few countries/territories where we do not have data, such as Pitcairn Island in below attribute table, but most of the features should now have values. Now we can develop a symbology to represent our newly joined data.

12. If Properties has been closed, reopen it, and click on the Symbology tab.

Currently the world-administrative-boundaries are represented as a single symbol. To create a choropleth map (also known as an aggregated thematic map or quantitative map), we select Graduated.

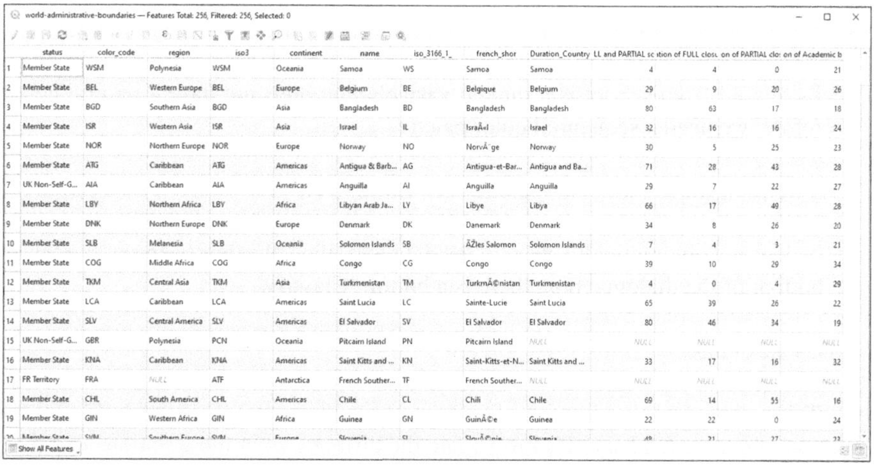
world-administrative-boundaries — Features Total: 256, Filtered: 256, Selected: 0

	status	color_code	region	iso3	continent	name	iso_3166_1_	french_shor	Duration_Country	LL and PARTIAL sc	ation of FULL clos	on of PARTIAL clo	on of Academic b
1	Member State	WSM	Polynesia	WSM	Oceania	Samoa	WS	Samoa	Samoa	4	4	0	21
2	Member State	BEL	Western Europe	BEL	Europe	Belgium	BE	Belgique	Belgium	29	9	20	26
3	Member State	BGD	Southern Asia	BGD	Asia	Bangladesh	BD	Bangladesh	Bangladesh	80	63	17	18
4	Member State	ISR	Western Asia	ISR	Asia	Israel	IL	IsraÃ«l	Israel	32	16	16	24
5	Member State	NOR	Northern Europe	NOR	Europe	Norway	NO	NorvÃ¨ge	Norway	30	5	25	23
6	Member State	ATG	Caribbean	ATG	Americas	Antigua & Barb...	AG	Antigua-et-Bar...	Antigua and Ba...	71	28	43	28
7	UK Non-Self-G...	AIA	Caribbean	AIA	Americas	Anguilla	AI	Anguilla	Anguilla	29	7	22	27
8	Member State	LBY	Northern Africa	LBY	Africa	Libyan Arab Ja...	LY	Libye	Libya	66	17	49	28
9	Member State	DNK	Northern Europe	DNK	Europe	Denmark	DK	Danemark	Denmark	34	8	26	20
10	Member State	SLB	Melanesia	SLB	Oceania	Solomon Islands	SB	ÃŽles Salomon	Solomon Islands	7	4	3	21
11	Member State	COG	Middle Africa	COG	Africa	Congo	CG	Congo	Congo	39	10	29	34
12	Member State	TKM	Central Asia	TKM	Asia	Turkmenistan	TM	TurkmÃ©nistan	Turkmenistan	4	0	4	29
13	Member State	LCA	Caribbean	LCA	Americas	Saint Lucia	LC	Sainte-Lucie	Saint Lucia	65	39	26	22
14	Member State	SLV	Central America	SLV	Americas	El Salvador	SV	El Salvador	El Salvador	80	46	34	19
15	UK Non-Self-G...	GBR	Polynesia	PCN	Oceania	Pitcairn Island	PN	Pitcairn Island	NULL	NULL	NULL	NULL	NULL
16	Member State	KNA	Caribbean	KNA	Americas	Saint Kitts and ...	KN	Saint-Kitts-et-N...	Saint Kitts and ...	33	17	16	32
17	FR Territory	FRA	NULL	ATF	Antarctica	French Souther...	TF	French Souther...	NULL	NULL	NULL	NULL	NULL
18	Member State	CHL	South America	CHL	Americas	Chile	CL	Chili	Chile	69	14	55	16
19	Member State	GIN	Western Africa	GIN	Africa	Guinea	GN	GuinÃ©e	Guinea	22	22	0	24
20	Member State	SVN	Southern Europe	SVN	Europe	Slovenia	SI	SlovÃ©nie	Slovenia	48	21	27	22

Show All Features

FIGURE 8.2 Screenshot of the attribute table with the completed join.

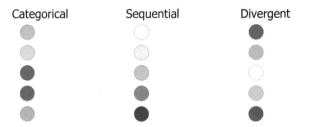

FIGURE 8.3 Example of a categorical, sequential, and divergent color palette.

13. Change Single Symbol to Graduated.
14. Select Duration of Full closures (in weeks) as the value.
15. We can select a range of different colors to represent this value. Click on the dropdown menu to explore this.

In the previous chapter, we selected the random color palette, which assigned a color to an attribute (e.g., type of litter) using a categorical or qualitative scheme. When working with categorical data, where there is no order to the ranking of values for that specific variable, such schemes are sufficient, although one must always consider selecting colors that maximize distinction and understanding, as well as considering those who are color blind. When working with variables whose values do have a ranking, it is important to make a distinction between selecting a sequential and diverging color scheme.

Figure 8.3 provides an overview of the three predominant color schemes used in GIS. A sequential scheme is used when our variable does not have a center, such as a variable describing rates or raw values. Color palettes are often designed and ordered using color lightness, but can also consider multiple color hues, which can provide another means by which differences can be visually discerned in the map. A diverging scheme is where our variable has a center, and the values trend away from each other, such as percentage change, which might center around zero (i.e., no change) with both negative and positive values. In our case study, the number of weeks missed from school is a ranked variable that does not have a natural center. Therefore, we need a sequential color palette.

16. Select a sequential color palette (i.e., Blues or Reds).

QGIS does not automatically classify the data once the value has been set, but toward the bottom of the dialogue box, there is a button 'Classify'.

17. Click Classify.

QGIS has classified the data into five bins of equal count, as shown in Figure 8.4. We can of course (and should) investigate the differences in these options and think about how many classes we might want. To create variation in our output, we might want more than five classes.

Choropleth Maps

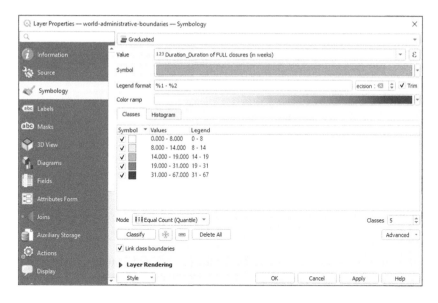

FIGURE 8.4 Screenshot of graduated symbology.

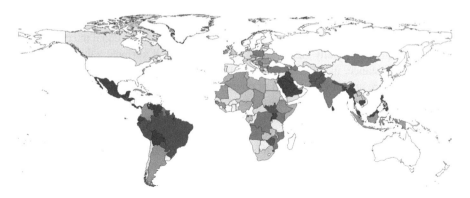

FIGURE 8.5 Screenshot of graduated symbology with ten classes classified using equal count.

18. In Classes, change five to ten.
19. For now, let us keep the Mode as Equal Count, click Apply (and close or move the dialogue box to see the map).

Our first instinct should be to look to discern patterns from the map. Your world map should resemble Figure 8.5, where we appear to have darker colors in the southern hemisphere, particularly South America, Africa, and east Asia. This classification has ensured that there are an equal number of countries in each class bin. It is important to consider that this output will change if we change the classification method.

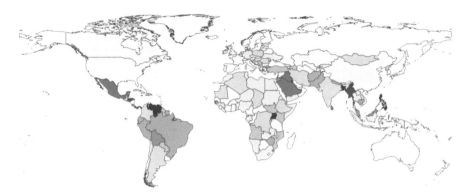

FIGURE 8.6 Screenshot of graduated symbology with ten classes classified using equal interval.

20. Return to the Properties and the Symbology tab and change the Mode from Equal Count to Equal Interval. Remember to click Classify.
21. Click Apply.

Your world map should now resemble Figure 8.6. Here we can see that the number of countries in the 'highest' band has reduced. The default, and the method that we first visualized this data using, is equal count (or quantile). This method divides the data into classes (or bins) so that each has an equal number of features. The equal interval method classifies data into equally sized ranges within the overall attribute values. For example, in this instance, the classification method has created ten bins of 6.7 weeks each. Therefore, there are only a handful of countries that appear to have been closed for the maximum amount of time, as opposed to the equal count method that visualized a lot more countries as being closed for longer. Neither method is incorrect, but it demonstrates the power of cartography and how the same data can be visualized with stark differences in interpretation, highlighting the importance of selecting an appropriate classifier.

There are also other methods of classification that we could use. Natural Breaks (Jenks) classifies the values into bins that appear naturally within the histogram of the data, grouping similar values together to maximize the difference between the classes. This is important, as such classification methods mean we cannot compare outputs of different data, as the breaks are unique to that dataset and cannot be compared objectively. Standard deviation classifies the data based on the proportion of the standard deviation in relation to the mean of the value. There are also two other classification schemes, logarithmic and pretty breaks.

22. Explore the different classification methods in visualizing this data.

There are also three other observations that can be made from visualizing this data. Firstly, some countries appear lower than we might expect (i.e., the USA). Secondly, we still have the issue of values straddling between classes (i.e., 0–6.7 and 6.7–13.4). Thirdly some administrative boundaries appear to be missing. Returning to the first point, the USA looks very low on this map, even though many

Choropleth Maps

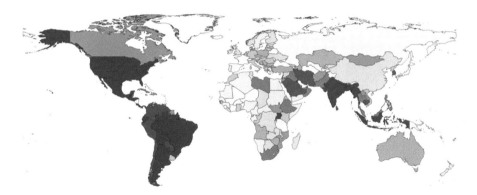

FIGURE 8.7 Screenshot of graduated symbology with ten classes classified using natural breaks for full and partial closures.

schools have been closed. This is because most of the information is contained in another attribute.

23. Click on the USA with the Identify Features button.

Here we can see that the number of FULL closures was 0. This is a function of how a FULL closure is defined by the data collection agency and is particularly relevant in countries which span large geographic areas. In these countries there are multiple state governments that control education, meaning the response may not have been uniform at the national scale. There was also a shift during the second and third waves of Covid-19 to enforce regional lockdowns, whereby not every school at a national level was closed. As such, we want to reconsider our visualization using the attribute that contains both FULL and PARTIAL closures.

24. Navigate back to Properties and set the value as Duration of FULL and PARTIAL school closures, with ten classes using Natural Breaks.

Our map should now resemble Figure 8.7. We can now see the higher impact of lockdowns on larger countries that did not necessarily have a uniform geographic policy whereby all national schools were closed, such as the USA and India.

Regarding the second point of values straddling boundaries, we again want to specify a 0.001 delineation to remove any ambiguity. This was explained in detail in the previous chapter.

25. Reopen the symbology tab.
26. Manually change the lowest classification value 0.000 to 0.001.

This should manually update the upper value, and your dialogue box should resemble Figure 8.8.

27. Click Apply.

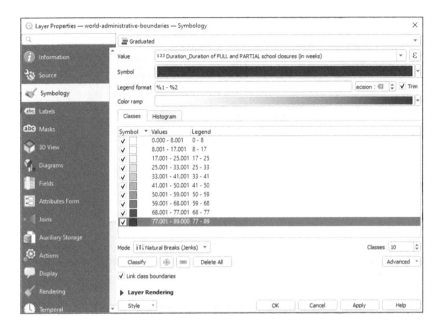

FIGURE 8.8   Screenshot of symbology tab to remove straddling of class boundaries.

We can see that there does not appear to be any change in the map output, suggesting that the values were not actually straddled; however, without clear instruction, particularly for cartography we should never take that risk.

Thirdly, if we were to zoom in, some of the smaller islands and territories are missing from the map. These locations have 'No Data', and thus are not being visualized in the map. This is because when visualizing data using the graduated symbology, null values are not included. We overcome this by changing the type of visualization.

28. Reopen the symbology tab in Properties and change the type of visualization from Graduate to Rule Based.

This will keep the existing symbology and classes that we have generated but provide us with the opportunity to present No Data.

29. Click on the Green + sign to add a rule.
30. Specify the label as 'No Data'.
31. Ensure that 'Else' is ticked.
32. Finally, change the color to a gray. This will be distinct enough from our current color palette, but not overpowering that it dominates the map.

Your dialogue box should resemble Figure 8.9.

33. Click OK to add the rule.

Choropleth Maps

FIGURE 8.9 Screenshot of parameters for the edit rule.

34. Click Apply and OK in the properties dialogue box.

Your map should now resemble Figure 8.10. We should now see locations appear on the map with a gray fill, including countries that we do not have information for, such as French Guiana in South America or disputed territories. It is not the aim of this book to discuss the geopolitics of disputed territories, therefore the categorization of countries was classified according to the data provider (World Food Programme 2019), with no pre-processing being undertaken. Therefore, the data is used 'as provided' and does not reflect any personal views or opinions.

Finally, we want to take our data that has been symbolized and generate a map in the Print Composer.

35. Create a new Layout called School Closures.
36. Add the map, this can cover almost the entire layout space.
37. Right click on the map, and open Item Properties.
38. Change the extent from −180 to 180 (X min to X max) and −80 to 90 (Y min to Y max), as shown in Figure 8.11.

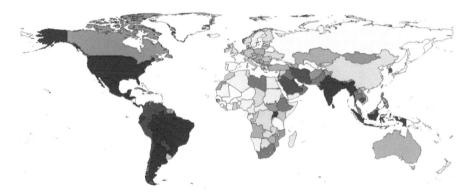

FIGURE 8.10 Screenshot of graduated symbology with ten classes classified using natural breaks for full and partial closures, with no data visualized.

```
▼ Extents
X min   -180.000
Y min   -80.000
X max   180.000
Y max   90.000
```

FIGURE 8.11 Screenshot to manually change extent in the print composer.

We touched upon this in the previous chapter. To overcome the trial and error of the map extent, where we know the exact coordinates we want to present the map at, we can manually insert these in the Print Composer using Item Properties. As this is a global map using latitude and longitude, we can specify these values accordingly. The only difference is that the minimum Y is provided as −80 instead of −90 to reduce empty space.

For maps that cover a global extent, a scale bar and north arrow are somewhat redundant. They will create noise in our map, and as navigation is not the main aspect of this work, we can keep them out of the final layout to ensure best presentation of the results. We do however need a legend.

39. Add a legend.
40. Change the legend name to 'School Closures (Weeks)'.
41. Ensure that the heading from 'Duration' is Hidden, by unchecking Auto Updates and right-clicking on the item.
42. Open the dropdown on world-administrative-boundaries (now School Closures (Weeks)) in legend items.
43. Manually change the legend so that the minimum value (with the exception of 0) is one higher to match our manual classification.

Choropleth Maps 143

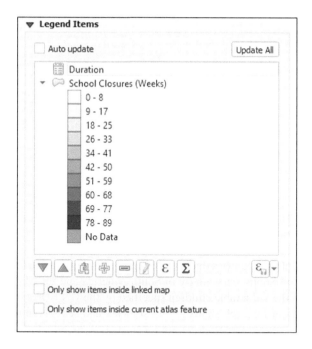

FIGURE 8.12 Screenshot of updated legend values.

Your legend items should resemble Figure 8.12. We should have a lot of space in the Pacific Ocean to place the legend. However, this will obscure some of the Pacific Islands.

44. In Item Properties for the Legend, untick Background.
45. To fit this in the map, we may need to reduce the font of the text. In Item Properties, in Fonts and Text Formatting, reduce the font size for the Legend Title.

Your final map should resemble Figure 8.13.

8.3 CASE STUDY CONCLUDING REMARKS

The Covid-19 global pandemic caused an unparalleled education crisis, with some countries having closed schools for almost the entire duration (up to when the data was downloaded in March 2022). Subsequently, knowing the locations where schools have been closed for the longest could support international efforts to provide additional support to the students and youth who have been impacted the most. This is not downplaying the challenges of students whose schools remained open during a pandemic, but when looking specifically through the lens of SDG4, such data and maps are an important tool. When this information is coupled with other indicators of education that highlight pre-Covid-19 education disparities of vulnerable children as highlighted by the UN SDGs, efforts can be directed toward providing national

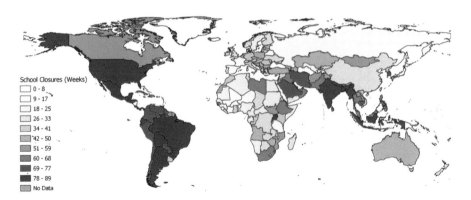

FIGURE 8.13 Thematic map of full and partial global school closures (in weeks) since the onset of the pandemic.

support. Many of the countries that are highlighted as being closed for a long duration are in the Global South, particularly in South America and the Malay Archipelago, where many of the vulnerable children specified by the UN SDGs live. There is also a growing body of research showing that students and schools in the Global South may not have access to remote learning tools, such as mobile devices and the internet, as readily as their peers in the Global North (Tomczyk et al. 2019; Johansen et al. 2021). Again, this is not to assume equal coverage in the Global North (see Chapter 5 for an example of this), but to highlight how choropleth maps can be used to visualize pre-existing disparities that may exacerbate the impacts of these closures.

In cartographic terms, this chapter has predominantly focused on the importance of exploring and selecting a suitable color palette, as well as identifying the methods to classify the information stored within the attribute tables. We have also reinforced the learning outcomes associated with the previous chapter, specifically navigating the Print Composer. In the next chapter, we continue to build on these skills.

8.3.1 TEST YOURSELF

To test yourself on the learning outcomes of this chapter, can you:

a. Generate a choropleth map of PARTIAL school closures at a global scale?

REFERENCES

Johansen, J., Noll, J., and Johansen, C., 2021. InfoInternet for education in the Global South: A study of applications enabled by free information-only internet access in technologically disadvantaged areas. *African Journal of Science, Technology, Innovation and Development*, pp. 1–13. https://doi.org/10.1080/20421338.2021.1884326

Tomczyk, Ł. and Sunday Oyelere, S., 2019. ICT for learning and inclusion in Latin America and Europe. *Case Study from Countries: Bolivia, Brazil, Cuba, Dominican Republic, Ecuador, Finland, Poland, Turkey, Uruguay*. Cracow: Pedagogical University of Cracow.

UNESCO (2022). *Total Duration of School Closures [dataset]*. Available from: https://data.humdata.org/dataset/global-school-closures-covid19. Accessed on March 1, 2022.

World Food Programme (2019). *World Administrative Boundaries – Countries and Territories [dataset]*. Available from: https://public.opendatasoft.com/explore/dataset/world-administrative-boundaries/information/. Accessed on March 1, 2022.

9 Change Maps

9.1 INTRODUCTION AND LEARNING OUTCOMES

In this chapter, we advance our cartography skills and develop another thematic map, a change map. Here we consider the importance of a divergent color palette in our symbology, as well as the methodology to generate a manual classification system to ensure that our color palette is consistent when representing values either side of a central value (i.e., zero). Finally, we create a map inset that provides a wider spatial representation of where the work is being undertaken to provide geographic context.

The learning outcomes are achieved in relation to SDG15.2. This target aims to promote the implementation of sustainable management of all forests, halt deforestation, and enhance afforestation. As such, we focus on how change maps can be developed using land cover data in France. In Europe, Corine Land Cover (CLC) data has been developed from Copernicus for the years 1990, 2000, 2006, 2012, and 2018. Subsequently, this data provides an excellent opportunity to explore locations of deforestation and afforestation at local, regional, national, and even continental scales. This is also a topic investigated regularly in my courses, whereby students want to use GIS to identify locations that are undergoing changes related to forest cover, due to the importance of such ecosystems for carbon regulation, biodiversity, wellbeing, and a range of other factors. By the end of this chapter, you will have completed three learning outcomes:

- Use the field calculator to quantify the change in forest cover
- Develop a manual classification scheme to visualize diverging data (i.e., percentage change)
- Use the Print Composer to insert multiple maps into a layout

9.2 CASE STUDY: SDG15.2 MAPPING THE CHANGING FORESTS OF FRANCE

1. Open QGIS and open the existing Ch9.qgz from the extracted zip folder.

There should be one hexagonal grid in the map canvas, as well as the world-administrative-boundaries file from the World Food Programme (2019) that we used in the previous chapter. The world-administrative-boundaries layer should be turned off.

2. Open the attribute table for the France_Grid layer.

Here we should see that there is an FID attribute, but also three area attributes (total area, area 2012, and area 2018). Total area represents the area of the hexagonal grid that the data is aggregated to. Area 2012 and area 2018 represent the total area of forest in that grid for 2012 and 2018 respectively.

The forest area was calculated from the CLC 2012 and CLC 2018 vector land cover layers downloaded from the EEA (2012, 2018). These datasets are licensed under the Copernicus data and information policy regulation (EU) No. 1159/2013 of 12 July 2013. This data is the sole property of the European Union (EU), which in no way endorses this work, but such work is only possible with funding by the EU. The layers have subsequently undergone pre-processing, and to ensure that this work can be transferred to other geographic regions or scaled up to a continental area, the details of this pre-processing are documented. Some of the geoprocessing terms (clip and intersect) are introduced in later chapters, but this should provide context upon completion of the book should you wish to revisit this. Firstly, the CLC 2012 and CLC 2018 vector land cover layers were downloaded and clipped to the outline of France. A new binary attribute using the field calculator representing forest cover was created from the attribute values 311, 312, and 313. Using this binary value, all features that were forest were selected and exported to their own feature layers for both 2012 and 2018. A hexagonal 5 km grid matching the extent of France was generated, which meant there were three layers, one representing the grid and two representing forest cover for the different time periods. Two separate intersects were then implemented to identify unique forest within each grid cell: one with the grid and Forest 2012 and one with the grid and Forest 2018. This created two new layers, which had the original FID value from the grid and a geometry attribute of the area. The attribute tables of the intersected forest layers were then joined back to the grid using the FID, and this is the layer we are working with.

One quick caveat before we begin. The minimum mapping unit of the CLC dataset is 25 ha, but these are complemented by a change map that has a minimum mapping unit of 5 ha. As such, the analysis we undertake has already been implemented by the EEA and presented as its own dataset at a finer spatial scale. Therefore, if we were interested in the changes between surveys for any localized policy or research, we should use the CLC-Change layers as these have a finer spatial resolution. However, as the rationale behind this chapter is to demonstrate how to model and map change using best cartographic principles, the chapter will work with the coarser data. This means we can apply this workflow to other national datasets in our own projects where this processing has not yet been completed.

Returning to the case study, we have two options here; we could map the change in total area, or we could map the percentage change in area. As the data is aggregated to a grid, we could report the total area as this should be somewhat standardized. However, it is good practice to present percentage change, especially as most geographic aggregation units are arbitrary in size and shape. Think back to some of the census data we have used, or even the previous chapter on countries, if we were to present such data on total forest, it could be skewed by geographic units that are larger. Also, we should know from Chapter 4 that while we have created equal grids, there will always be distortion in our projections and reporting percentage change is the best option.

3. Click on the Field Calculator button in the attribute table.
4. Create a new variable called Percent12.
5. Set the type to Decimal Number.

Change Maps

FIGURE 9.1 Screenshot of parameters needed for the field calculator.

6. Set the expression as: ("Area2012" / "Area") * 100.
7. Ensure your expression resembles Figure 9.1. Click OK.

By now, we should always check that our commands have successfully completed.

8. Check the attribute table to see if this has been completed correctly. If not, return to the field calculator and check the expression.
9. Repeat this process to generate a new Percent18 field using the 2018 values.

We now have two additional variables in the Attribute Table: Percent12 and Percent18. These attributes represent the percentage of forest cover within each hexagonal unit for the 2 years. Next, we want to combine them to identify how much change occurred within each unit.

10. Click on the Field Calculator button.
11. Create another new variable called Change.
12. Set the type to Decimal Number.
13. Set the expression as: "Percent18" – "Percent12".
14. Click OK.

This should create a new variable, with negative values representing a decrease in forest cover from 2012 to 2018 and a positive value representing an increase in forest cover. For example, if there was more forest cover in 2012 (in other words deforestation), then subtracting forest cover in 2012 from 2018 would result in a negative value. Explore the data in this column by sorting it. The values range from −27.4 to 69.1. This range is interesting, but it does not tell us about the quantity or where these changes occur.

15. Save the attribute table edits using the Save Edits button, and turn off Toggle Editing.
16. Open Properties for the grid and navigate to Symbology.
17. Change the type from Single Symbol to Graduated.
18. Select Change as the value.
19. Keep the color ramp as default for now, we explore that in detail in a moment.
20. We can also keep the default classification mode and number of classes.
21. Click Classify.
22. Click Apply.

Your map should resemble Figure 9.2; however, at this stage it is not the most intuitive. We should consult the classification table in Properties (this is visible in Figure 9.2).

The equal count classification mode has created five classification bins, but three of them represent 0-0, and the first represents −27–0. This means that most of our

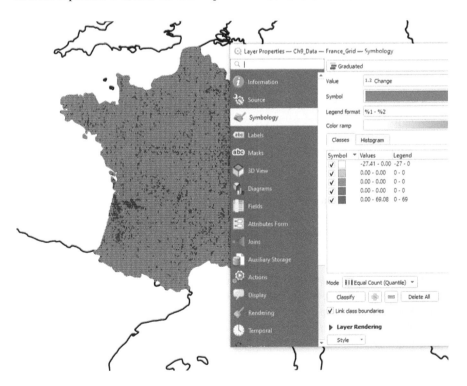

FIGURE 9.2 Screenshot of initial symbology in properties and applied on the data.

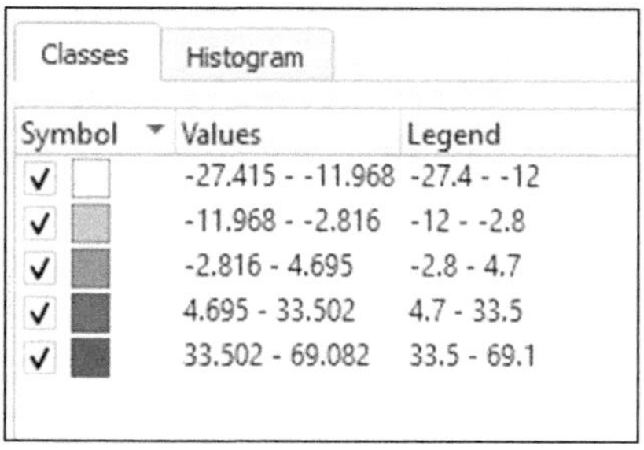

Change Maps

Symbol	Values	Legend
✓	-27.415 - -11.968	-27.4 - -12
✓	-11.968 - -2.816	-12 - -2.8
✓	-2.816 - 4.695	-2.8 - 4.7
✓	4.695 - 33.502	4.7 - 33.5
✓	33.502 - 69.082	33.5 - 69.1

FIGURE 9.3 Classification scheme of five classes using Natural Breaks.

data is 0 (no change) but is not currently visualized as the layers render in the order they are listed. Therefore, all areas of deforestation and no change are visualized as white. If we change the classification mode to Natural Breaks, we should see a difference.

23. Open Properties if it closed and navigate to Symbology.
24. Change the Classification mode to Natural Breaks.

Your classification should now resemble Figure 9.3. We now have variation in our symbology, but we need to think how meaningful this representation is. We are currently using a sequential color palette, but the values range from negative to positive values. Therefore, we would ideally represent the values using a diverging color palette, with a gradient of colors to represent negative values and another to represent positive values.

25. In the symbology tab, change the color palette to the diverging color scheme RdGy.
26. Click Apply.

Your screen should now resemble Figure 9.4. This should begin to look a bit more informative, with red values representing negative changes and black values positive ones. However, the neutral color representing no change contains values that range from −2.8 to 4.7, meaning it includes values that have undergone deforestation, afforestation, and no change. To overcome this, we can specify our own classification scheme.

27. In the symbology tab, change the mode to equal interval.

Because our values range from −30 to 70, we can create bins of 10%.

28. Specify 11 classes.
29. Next, we want to manually change the values to match our 10% bins. In the dialogue box, double click on −27.414− −18.642 under the values

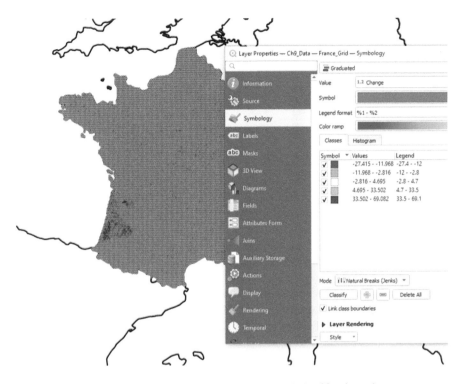

FIGURE 9.4 Screenshot of the diverging color palette and classification scheme.

This should open another dialogue box, where we can change the maximum and minimum values.

30. Change the lower value to −30 and the upper value to −20.001.
31. Click OK.
32. Repeat this process for all the bins. The only exceptions are for the 0 bin, which represents no change. This should simply read −0.001 to 0.001. While this might also include locations which have slightly less than 0.001% or −0.001% change, there are no values in this dataset. Similarly, when we reach positive numbers, the lower value should read 0.001 to 10.
33. Ensure the classification scheme resembles Figure 9.5. Click Apply.

The good news is that we have a diverging color scheme; however, this is not centered on 0. We could manually change the colors, but this would result in colors of the same lightness representing different quantities (i.e., the darkest gray representing 60%–70% while the darkest red would only represent −20% to −30%). Therefore, we want to ensure that we have the same number of classes in the positive and negative bins. While the values do not go to 100% change, it may still be useful to represent this. Therefore we will present the results from −100% to 100%. Therefore, we actually want to create 21 classes.

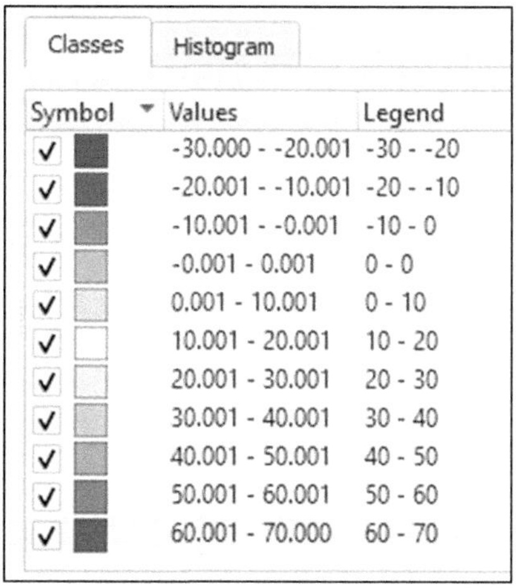

Change Maps 153

Symbol	Values	Legend
✓ ▇	-30.000 - -20.001	-30 - -20
✓ ▇	-20.001 - -10.001	-20 - -10
✓ ▇	-10.001 - -0.001	-10 - 0
✓ ▇	-0.001 - 0.001	0 - 0
✓ ▇	0.001 - 10.001	0 - 10
✓ ☐	10.001 - 20.001	10 - 20
✓ ☐	20.001 - 30.001	20 - 30
✓ ▇	30.001 - 40.001	30 - 40
✓ ▇	40.001 - 50.001	40 - 50
✓ ▇	50.001 - 60.001	50 - 60
✓ ▇	60.001 - 70.000	60 - 70

FIGURE 9.5 Screenshot of the manually changed classification scheme.

34. Repeat the above steps but use 21 classes, starting at −100 and ending at 100.

Your classification scheme should now resemble Figure 9.6. This repetition in manually changing the values should have reinforced the learning outcomes associated with straddled values. Upon completion of this, we now have a complete diverging color palette, with equal color lightness for values of equivalent sizes but different signs. Now we must consider the contrast in colors. Because our values do not span the full range, with most percentage changes occurring at less than 20%, our colors may appear washed out in a final map. Therefore, we want to edit the existing color palette to ensure all grids are readable.

35. On the drop-down menu for color palette in the Symbologies tab, click Edit Color Ramp.

Here we have more advanced options, showing the range of colors that are used within the palette. We can see that black is not actually black, but a dark gray. Let's change that.

36. Change Color 2 to Black.

Next, we want to change the gradient position for the mid-points. This will reduce the number of light colors for lower values.

37. Slide the second position from 25% to 48% and the fourth position from 75% to 52%, as shown in Figure 9.7.
38. Click OK, and then apply the color palette.

This process will remove some of the variation among afforested areas and deforested areas, but it will substantially enhance readability at a national scale. For

Symbol	Values	Legend
✓	-100.000 - -90....	-100 - -90
✓	-90.001 - -80.001	-90 - -80
✓	-80.001 - -70.001	-80 - -70
✓	-70.001 - -60.001	-70 - -60
✓	-60.001 - -50.001	-60 - -50
✓	-50.001 - -40.001	-50 - -40
✓	-40.001 - -30.001	-40 - -30
✓	-30.001 - -20.001	-30 - -20
✓	-20.001 - -10.001	-20 - -10
✓	-10.001 - -0.001	-10 - 0
✓	-0.001 - 0.001	0 - 0
✓	0.001 - 10.001	0 - 10
✓	10.001 - 20.001	10 - 20
✓	20.001 - 30.001	20 - 30
✓	30.001 - 40.001	30 - 40
✓	40.001 - 50.001	40 - 50
✓	50.001 - 60.001	50 - 60
✓	60.001 - 70.001	60 - 70
✓	70.001 - 80.001	70 - 80
✓	80.001 - 90.001	80 - 90
✓	90.001 - 100.000	90 - 100

FIGURE 9.6 Screenshot of the manually changed classification scheme from −100 to 100.

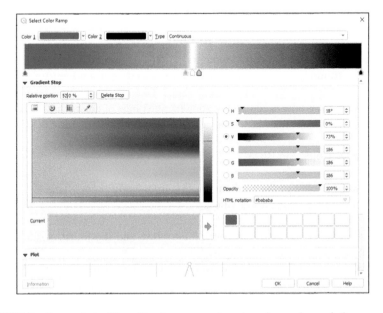

FIGURE 9.7 Screenshot of the edit color ramp options to reduce color variation.

Change Maps 155

datasets where the values range from −100 to 100, this may not be necessary, but for data which is predominantly grouped around the central value, it provides contrast in the map output for the end user.

However, despite these changes, the map is still not as easily interpreted. This is because each grid has an outline. To improve our map, we should either remove the borders or change them to the same color as the fill.

39. Double click on one of the symbols (colored square) in the classification table on the Symbologies tab.

This should open a dialogue box, where we can manually change the color and scheme.

40. Click on Simple Fill, which again should change the dialogue box.
41. Click on the drop-down menu of the Fill Color, and select Copy Color.
42. Click on the drop-down menus of the stroke color, and select Paste Color – this should update the color to match the fill.
43. Click OK.
44. Repeat this for each of the 21 colored squares in the classification table.

This can be quite a finicky and arduous process; no-one said making good maps was simple… However, should something happen, we do not necessarily want to go through this process again. Therefore, we can save this color palette, and re-use it in the future.

45. Click on Style at the bottom of the dialogue box.
46. Select Save Style.
47. Save this as a QGIS QML Style File named CoverChange in your working directory.

This is demonstrated in Figure 9.8.

Should we accidentally change the symbology, we can click on Style > Load Style and select this file that we have just saved, and it will load.

48. Apply this symbology.

We now have a well-contrasted map, as shown in Figure 9.9, with a clear depiction of the border, as well as where we observe areas of deforestation and afforestation between 2012 and 2018. If your map appears slightly different, you may need to reorder the layers in the Layer Panel. Now it is time to make our map in the Print Composer.

49. Open a new Print Composer.
50. Insert the map, ensuring that the data is suitably zoomed in. Make sure that this map takes up just over half of the layout page on the right-hand side, leaving space for a legend on the right of the map. We will be inserting additional maps, so we want to keep the space free.

FIGURE 9.8 Screenshot of the save layer style dialogue box.

FIGURE 9.9 Screenshot of custom color palette applied to the cover change attribute.

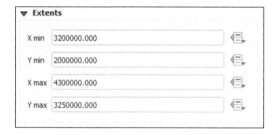

FIGURE 9.10   Screenshot of optimal coordinate values.

If we have difficulty capturing the zoom perfectly, we can specify the extent like we did in the previous chapter. Try using the coordinate information at the bottom of the map canvas to see if you can identify the maximum and minimum XY values.

51. Change the extent values to best place France in the layout (suggested values provided in Figure 9.10, but make sure you try and identify these yourself).
52. Add a scale bar.
53. Add a north arrow.
54. Add a legend to the right of the map.

We may want to clean up our legend a bit. In the item properties, we can manually change the labels. Because cartography is about presenting the information as effective as possible, it is perhaps unnecessary to report every color value. Therefore, we can remove most of the labels to report simply Increase (10% bins), No Change, and Decrease (10% bins).

55. Change −70 to −60.001 to Decrease (10% bins).
56. Change −0.001 to 0.001 to No Change.
57. Change 60.001–70.001 to Increase (10% bins).
58. Change all other labels to a single space to remove the text.

The placement of Increase and Decrease at the 60%–70% bin was to provide spacing in the legend so that it was not too squashed at the top and bottom of the item. Next, we need to remove the unnecessary items from the legend, which consists of the world-administrative-boundaries layer.

59. Select the world-administrative-boundaries in the legend items panel and click on the red minus button to remove it.
60. Change the name of the legend to 'Forest Change'.
61. Make sure the layer name is hidden (by right clicking it in item properties).

Make sure the layout is set up in a manner that provides space for two additional maps, as shown in Figure 9.11. If we add more layers to our map document, it will update in the Print Composer. Therefore, we need to lock this in place.

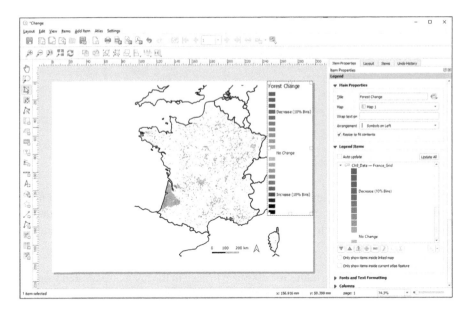

FIGURE 9.11   Screenshot of map location in Print Composer.

62. Select the map in the Print Composer and open Item Properties.
63. Under Layers, ensure that Lock Layers and Lock Styles for Layers are ticked.

These options will ensure that the presentation does not change when we return to the map canvas and start adding new layers.

64. Return to the map canvas and turn off the France_Grid layer, leaving only the world-administrative-boundaries layer.

This is the map of the world. We are interested in highlighting where France is in the context of Europe.

65. Select France using the Select Tool.
66. Right click on the world-administrative-boundaries layer in the Layers Panel, and export the selected feature to its own layer in the Ch9 GeoPackage called France.
67. Change the symbology of France to a purple color (and increase the stroke width to 1) and change the world-administrative-boundaries to a green color, using Single Symbol. Note this has to be completed for both layers separately.
68. Zoom to an overall extent of Europe.
69. Return to the Print Composer, and add this new map in the top left-hand corner.

This new inset map now references the location of France within the wider extent of Europe. Such maps are important in the global context of our work. Creating maps

Change Maps 159

of specific locations is important, but we cannot always assume that the reader of the map will know exactly where the location of interest is. While one would hope that everyone is somewhat familiar with the geographic location of France, this is not always the case. We are also providing the reader with a consideration of where France is situated within Europe, and from the inset map it is clear that the eastern coast is exposed to the Atlantic Ocean (this will become important when analyzing the final map).

Finally, the area in the south-west of France looks like it has had a lot of deforestation and afforestation activity. Therefore, we might want to zoom in and create another inset map of this location.

70. Lock the layers for the European context map in the Item Properties.
71. Ensure that Frame is ticked for the European context map to clarify that it is an additional map and not a continuation of the main map.
72. Return to the map canvas, deselect the world-administrative-boundaries layer, and change the symbology back to a transparent color with a black border.
73. Turn off the France layer and turn back on the France_Grid layer.
74. Zoom to the area of south-west France shown in Figure 9.12.
75. Return to the Print Composer and add a third map in the bottom left of the layout. This should be the zoomed in area.

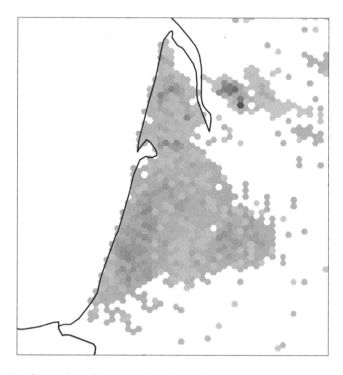

FIGURE 9.12 Screenshot of the area of south-west France.

76. Click on the first map (Map 1), which is the whole of France.
77. In the Item Properties, scroll to Overviews.
78. Click on the Green + to add an overview.
79. Set the Map Frame to Map 3 (this is the one we have just added).

The area that is represented in Map 3 should now be highlighted in the first map.

80. Change the symbology to a transparent fill and a solid red line for the outline, with a stroke width of 0.5. To do this easily, you will need to click on the color to open up symbol settings and then click on Simple Fill.

While the outline is obvious to us, we may want to add some lines and a Frame to connect the two maps.

81. Click on Map 3, and in Item Properties scroll down to Frame.
82. Ensure that it is ticked, and change the color to red with a thickness of 0.5.
83. In the vertical Layout toolbar, select Add Arrow.
84. Draw an arrow between the top right-hand corners of the two matching frames (remember to right click to finish drawing).
85. Click on the item properties for the arrows (called Polyline in the list of Items on the right of the screen).
86. Change the arrow to none for the head.
87. Change the color to match red.

Finally, to aid in interpretation, we should label our three maps a, b, and c.

88. Click the Add Label button (three times), and add a), b), and c) for the Europe map, France map, and south-west map, respectively.

Your map should hopefully resemble that presented in Figure 9.13.

Before we save our final map, we have highlighted the importance of selecting suitable color palettes, particularly to support color-blindness. QGIS now has functionality to investigate how colors are differentiated using different simulates of color-blindness.

89. Navigate to the tab View > Preview in the Print Composer.
90. Select the option Simulate Achromatopsia Color Blindness.

Your map should now be presented in grayscale. You can explore how this map looks using simulations of different colors. In all instances, this map output is readable, meaning we have selected a suitable set of colors, but this is also a good habit to begin to use in order to increase inclusivity and readability in cartography.

91. Save the map as an image.

Change Maps 161

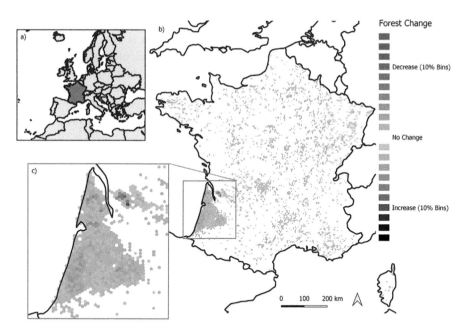

FIGURE 9.13 Percentage of forest cover change in France between 2012 and 2018. Image (a) provides an overview of the position of France within Europe, (b) provides the overall forest cover change at a national level, and (c) is an inset map of the south-west of the country where there is a lot of change.

9.3 CASE STUDY CONCLUDING REMARKS

We have now created a professional looking map that situates France within the European context and highlights an area that appears to have undergone an extensive amount of deforestation and afforestation since 2012. One thing that is perhaps missing from this map is a set of labels representing cities so that we can orientate ourselves to the largest settlements in the country; however, in its absence, the area we have zoomed on in Panel c is Bordeaux. In 2009, this region suffered a large amount of deforestation due to cyclone Klaus. Winds reached upward of 160 km/h (or 100 mph) and caused extensive deforestation due to the shallow depth of the root systems (Planque et al. 2017). Almost a decade later in 2018, the changes observed represent areas that have begun to afforest, through plantation and natural succession. The areas of deforestation represent areas that continue to decline in tree cover, most likely as a result of initial deforestation, with damaged trees continuing to be cleared, or private areas where landowners have not replanted. For those people not familiar with France, the context map panel a) highlights the geographic location of the west coast of France, where it can be seen that this area is coastal in nature and would be subject to cyclones from the Atlantic Ocean, highlighting the importance of these types of maps.

While this location represents an extreme example of deforestation, such maps provide national context to highlight the extent of afforestation. Importantly, it highlights that this is the largest expanse of forest area that is changing, but there are smaller areas throughout the country where deforestation is also occurring. These values represent approximately a loss of forest within 20% of existing 2012 cover. By mapping the changes, resources can be directed toward identifying the causes of these conversions, as well as identifying targeted policy measures to support possible private landowners increasing afforestation efforts within the country.

In this chapter, we built upon the options available to us in the Print Composer, identifying how to add multiple maps to a layout, and provide linkages between these maps. We have also further explored the symbology options, in particular how to create our own manual classification, that navigates the unique conditions of pivoting around a central value. We also saved these classification schemes, allowing us to re-use such symbologies across different maps, ensuring a consistent visualization to aid interpretation. In the next chapter, we create dynamic visualizations from our data to present our maps in a different format.

9.3.1 Test Yourself

If you want to test yourself on the learning outcomes of this chapter, complete the following:

a. Generate another inset panel that highlights an alternative area of key deforestation and/or afforestation.

REFERENCES

EEA and Corine Land Cover (CLC) (2012). *Version V2020_20u1*. Release Date: 13-05-2020. European Environment Agency. https://land.copernicus.eu/pan-european/corine-land-cover/clc-2012?tab=mapview.

EEA and Corine Land Cover (CLC) (2018). *Version 20b2*. Release Date: 21-12-2018. European Environment Agency. https://land.copernicus.eu/pan-european/corine-land-cover/clc2018.

Planque, C., Carrer, D., and Roujean, J.L., 2017. Analysis of MODIS albedo changes over steady woody covers in France during the period of 2001–2013. *Remote Sensing of Environment*, 191, pp. 13–29.

World Food Programme (2019). *World Administrative Boundaries – Countries and Territories [dataset]*. Available from: https://public.opendatasoft.com/explore/dataset/world-administrative-boundaries/information/. Accessed March 1, 2022.

10 Dynamic Visualization

10.1 INTRODUCTION AND LEARNING OUTCOMES

As we have now established through this section, cartography is essential for visualizing geographic phenomena and is an important tool in the cognitive processing of data, providing the link between humans and computers. The maps created thus far represent traditional cartographic methods, whereby geographic information represented as points, lines, and polygons are depicted on a map. This cartographic communication paradigm within GIS has resulted in the geographic phenomena we have been presenting as a series of 'snapshot' views (Laube 2014). For example, in previous chapters, we have visualized plastic waste, education access, and land cover on yearly and decadal timeframes, even though such features are continually moving and/or changing. This relatively static representation can limit our understanding of the geographical processes we are investigating (Dodge 2016), and within QGIS there is the possibility to visualize the temporal components of the spatial data.

Recent advances in technology, such as GPS, have resulted in an unprecedented amount of spatiotemporal geographic information. The quality and availability of these spatiotemporal data are allowing researchers to ask increasingly unique and novel questions, related to a range of topics (Holloway & Miller 2018). The use of road-sensors, taxi-networks, and mobile phones allows organizations and companies to track real-time traffic, which can be visualized on maps to allow users to make informed decisions regarding possible navigation routes. Such information is useful in our everyday lives, but can also be linked with SDG11 Sustainable Cities. Similarly, GPS data is perhaps most synonymous with animal tracking, with individual animals tracked across countries, continents, and even hemispheres. Such information is fundamental to conservation efforts, actively supporting SDG14 Life Below Water and SDG15 Life on Land, as barriers to key life cycle events, such as migration, can be identified. In this chapter, we explicitly focus on SDG15.5, which aims to halt the loss of biodiversity. Here, we investigate the hourly and quarter-hourly movement steps of zebras in Botswana, using GPS data from Bartlam-Brooks et al. (2013), obtained via Movebank (Bartlam-Brooks & Harris 2013) which have been licensed under CC0 1.0 Universal (CC0 1.0) Public Domain Dedication. Movebank (https://www.movebank.org) is an online geospatial repository, where researchers can store, manage, and analyze animal trajectory data. By the end of this chapter, you will have completed three learning outcomes:

- Appreciate the emerging importance of dynamic tools in understanding the physical world
- Use embedded dynamic functions to represent both the spatial and temporal elements of geographic data
- Explore the use of multimedia to create engaging cartographic products that can be used for science outreach

FIGURE 10.1 Burchill's zebra's in Makgadikgadi National Park, Botswana, 2014. (Photo credit: Paul Holloway.)

10.2 CASE STUDY: SDG15.5 A YEAR IN THE LIFE OF A ZEBRA

With the recent removal of a veterinary fence in Botswana, an historic zebra migration was recorded in Botswana from the Okavango Delta to Makgadikgadi National Park for the first time since anecdotal accounts in the early 20th century (Figure 10.1). Confirmation of this migration event was made possible by GPS tracking devices such as those attached to the animals in the study by Bartlam-Brooks et al. (2013). GPS locations of animals are measured at regular intervals (e.g., every 1 hour), with the location of these points representing the GPS fix. The data has already been downloaded and pre-processed to remove locations with no coordinate or time data associated with it from Movebank, but should you wish to explore the website, please do so.

1. Open a New Empty Project in QGIS.
2. Add the zebra data in the GeoPackage found in the extract Ch10 Data folder by navigating to tab Layer > Add Layer > Add Vector Layer.

To ensure the data is in the correct place geographically, add a basemap. For visualization purposes, a satellite basemap is more suited to the task of this chapter.

Dynamic Visualization

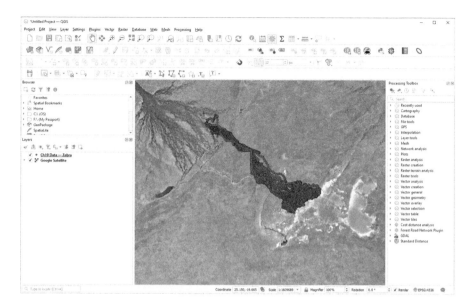

FIGURE 10.2 Screenshot of point data representing zebra locations in Botswana. Basemap is Google Satellite which is Map data ©2015 Google, please see https://www.google.at/permissions/geoguidelines/attr-guide.html.

3. Click on WebMapServices, select Search QMS, and add a satellite basemap of your choosing. Again, the instructions provided herein are visualized with Google Satellite.

Note Figure 10.2 is slightly zoomed out to capture the Okovango Delta in the northwest corner of the image and the Makgadikgadi Salt Pans in the south-east corner. The points representing zebra locations quite clearly show the location of zebras during the migration between the two locations. The dataset contains movement on seven zebras. The first thing we can do is to distinguish between these individuals using a simplistic color palette as we did in Chapter 7.

4. Right click on Properties and open the symbology tab.
5. Change Single Symbol to Categorized.
6. Specify the Value to 'tag-local-identifier'.
7. Keep the color ramp as 'random colors'.
8. Click Classify at the bottom of the dialogue box.

Your dialogue box should resemble Figure 10.3 (albeit with different random colors). There are seven different values, each representing an individual zebra. There is also a symbol depicting points that have not been classified.

9. Click OK.

While displaying this point data is an effective starting point for visual exploration, due to the unprecedented amounts of movement data that are being collected, the display is somewhat cluttered with an excessive number of data points. Dynamic

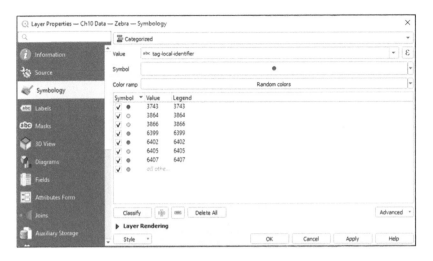

FIGURE 10.3 Screenshot of parameters for the symbology tab.

visualization is imperative to successful exploration and analysis of the data, as we can observe the daily movements of animals as never before seen. However, we have yet to complete the second check of our data, which is the attribute table.

10. Open the attribute table and navigate to the datetime column in the far right of the table.

We are going to use this datetime field to create a dynamic output of the zebra movements. However, we first need this to be recognized as a date and time field. We should check this in the Fields tab of Properties.

11. Open Properties for the Zebra layer.
12. Navigate to the Fields tab and scroll down to datetime.

This field type is specified as a DateTime factor, meaning we have the necessary data format to continue with the dynamic presentation.

13. In the Properties dialogue box, navigate to the Temporal tab.
14. Tick 'Dynamic Temporal Control'.
15. Change the configuration to 'Single Field with Date/Time'.
16. Ensure that datetime is the field (as this is the only date/time field in the attribute table, it should be the only variable returned).
17. Finally change Event Duration to 1 hour.
18. We have the option to accumulate features over time. What this means is that after every 1-hour time window, the previous locations will either disappear or accumulate. For the moment, let us leave this unticked.

Your dialogue box should resemble Figure 10.4.

Dynamic Visualization

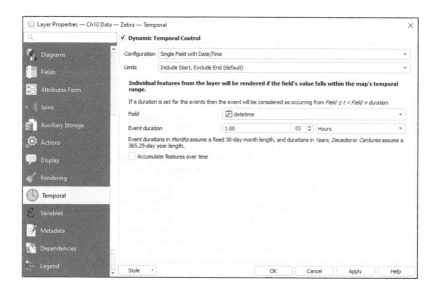

FIGURE 10.4 Screenshot of the parameters needed for the Temporal tab in properties.

19. Click OK.

While we have enabled the temporal settings in the layer, we need to switch on the relevant tools and toolbar affiliated with it.

20. Right click on the gray area to the right of the toolbars (as outlined in Chapter 2) and turn on the Temporal Controller Panel.

Your screen should now look similar to that portrayed in Figure 10.5. Above the map, the temporal controller has appeared; however, the tool itself is disabled. We must enable it.

21. Click the Animated Temporal Navigation button.

A slider appears that has play, pause, and skip buttons. We can also specify the timeframe of the animation range. The default of this function is to set this period for exactly 1 week prior to the current time, which does not align with when the data was collected. As such, there should not be any points in the display. Based on the skills in Chapters 4–7, we should be able to independently identify the earliest and latest dates of GPS fixes using the attribute table. We are reaching the point in this book where it is good practice to try and complete some of the fundamental GIS tasks without step-by-step instructions. These will be provided below but try to complete them independently if possible.

To identify the time range of GPS fixes.

22. Open the attribute table.

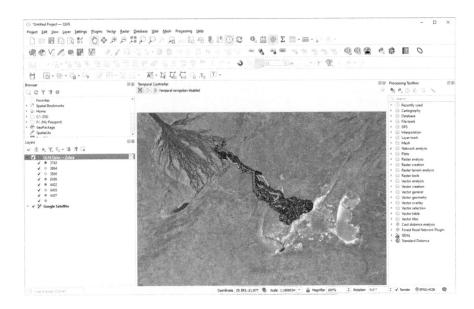

FIGURE 10.5 Screenshot of QGIS with the temporal controller turned on. This is visible above the map. Basemap is Google Satellite which is Map data ©2015 Google, please see https://www.google.at/permissions/geoguidelines/attr-guide.html.

FIGURE 10.6 Screenshot of temporal controller specifications.

23. Navigate to datetime field.
24. Click on the column to display it in ascending order, and note the earliest date (25 October 2007).
25. Click on the column again to display it in descending order and note the latest date (1 June 2009).

Now, we return to the dynamic temporal sequence.

26. Type the date (and time) acquired from the attribute table. If you have not noted the times, just simply use 09:00.
27. Specify the step as 1 hour.
28. Make sure the slider is placed on the left.

Ensure the Temporal Controller matches that shown in Figure 10.6. We should also only have a handful of points visualized as we are now viewing locations at 9 am on 25 October 2007 only.

Dynamic Visualization

29. Click the play button.

The points should begin to move as the time progresses. Our task is to engage with the dynamic visualization. In each hour, there should either be 1 or 4 points reflecting the hourly or quarter-hourly movements that have been collected. A good starting point is to try and describe the landscape and movement patterns we are observing.

Given the timeframe of this dataset (~18 months) this could be quite a long process to watch. Therefore, in the temporal controller, change the step to 1 day. This will still record all the fixes, but over a 24-hour period, which may make it easier to spot environmental interactions.

30. Change the step from 1 hour to 1 day.

We begin to see different patterns, with long linear movement steps as well as smaller daily movement paths that appear clustered. These patterns are indicative of migrating and foraging, respectively.

For example, Figure 10.7 is a 24-hour period for zebra 6399 in November 2008. We can quite clearly see the linear feature of its movement and if we pause the

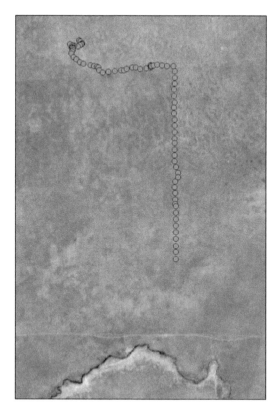

FIGURE 10.7 Screenshot of Zebra 6399 in November 2008. Basemap is Google Satellite which is Map data ©2015 Google, please see https://www.google.at/permissions/geoguidelines/attr-guide.html.

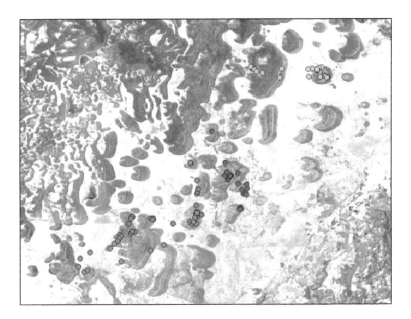

FIGURE 10.8 Screenshot of zebras appearing to avoid the salt pans in Makgadikgadi National Park. Basemap is Google Satellite which is Map data ©2015 Google, please see https://www.google.at/permissions/geoguidelines/attr-guide.html.

controller and zoom into the landscape and observe the basemap, we might be able to discern what feature is causing this. It appears to be a linear path, but it is actually a veterinary fence, with a fire break either side of it, meaning it provides a linear structure in which the individuals can move but cannot cross.

Similarly, if we observe the zebras in Makgadikgadi National Park in Figure 10.8, very few of the individual points fall directly on the salt pans. This suggests that the zebras are avoiding these locations, possibly due to exposure to the elements, predators, or lack of food.

As we can see in the above images, the patterns and processes appear somewhat evident, but the static nature of these screenshots is simply that, static. QGIS provides us with the opportunity to generate more dynamic visualizations of these outputs in the form of processed snapshots, and then videos. For this section of the chapter, we focus on the north-south migration of November 2008.

31. Change the settings on the slider to start on 1 November 2008 and end on 30 November 2008.
32. Change the time steps from 1 day (or 1 hour) to 4 hours.

Your Temporal Controller should resemble Figure 10.9.

33. Click the save button on the temporal controller, which is called the 'Export Animation' button.

Dynamic Visualization

FIGURE 10.9 Screenshot of the parameters for the temporal controller dates.

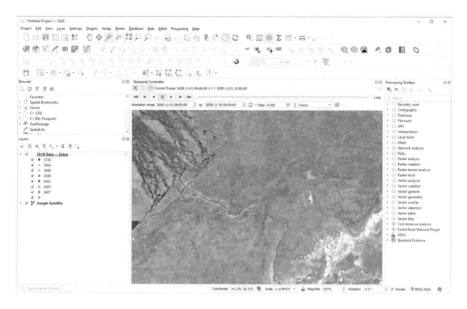

FIGURE 10.10 Screenshot of a suitable map extent for use in the animation including both the Okavango delta in the north-west and Makgadikgadi Salt Pans in the south-east. Basemap is Google Satellite which is Map data ©2015 Google, please see https://www.google.at/permissions/geoguidelines/attr-guide.html.

A dialogue box should open.

34. Set the output directory to a new folder called Chapter10_Images in your working directory. Note that this folder must be manually created.
35. Choose the Map Canvas Extent, ensuring your existing extent includes both the delta and the pans. This will set the output width and height. For reference, aim for a similar extent as Figure 10.10.
36. In case the location is accidently changed, click the padlock to lock these specifics in.
37. Ensure your dialogue box resembles Figure 10.11. Click Save.

The animation will run in the background as the PNGs are being saved. This may take a couple of minutes to fully process. Once it reaches 100%, we convert those PNGs to a video. Again, the instructions will be for Windows.

38. Navigate to the folder the files were saved in.

FIGURE 10.11 Screenshot of the parameters for the export map animation settings.

If we click on the link that appears at the top of the QGIS map canvas upon completion, it will take us right there. We should have 174 files.

39. Double click the first photo to open it in Photos (this should be called 0000).
40. Click on 'Edit & Create' and select Create a Video with Music. This will either be a main option, or located in options as Edit More.
41. Name the video something intuitive, such as Zebra Migration.

If working in Windows, you should have an interface that resembles Figure 10.12.

42. In the Project Library, click Add > From this PC.
43. Navigate to the folder and add all the additional files in the folder, but not 0000. If we hold down shift, it should select them all in one go. Once we have selected them all, click Open.
44. They should load into the library and be selected automatically. Click on Place in Storyboard.

These should now be added to the bottom of the screen, with a 3.0 in white written on it. This is the duration of the frame; however, this might be a bit long for our purposes.

45. Click on the first image so it is selected, then select all the rest by holding 'Ctrl + A'.
46. Click on Duration, and change the time to 1 second.

Dynamic Visualization

FIGURE 10.12 Screenshot of the Create a Video tool in Windows. Basemap is Google Satellite which is Map data ©2015 Google, please see https://www.google.at/permissions/geoguidelines/attr-guide.html.

47. In the top right-hand corner, click 'Finish Video'.
48. If the device has space, choose the high-quality value, otherwise choose a lower quality value.
49. Save the file in your working directory.

In this instance we have not added music, but should you wish, please do so. Once this has exported, we can watch it.

10.3 CASE STUDY CONCLUDING REMARKS

By watching the video, we should immediately observe the benefits to our spatial cognition. Visualization allows the identification of patterns that may not be evident across multiple spatial and temporal scales, as well as supporting hypothesis generation. This is evident in the above case study, as we can explore movement patterns in a dynamic fashion, positing explanations for faster, slower, or unusual behavior. For example, in the video we can observe that zebras are covering a much larger distance when moving along linear features, which we may hypothesize is down to the desire to avoid predation or human-wildlife conflict. Some of my previous research using telemetry datasets (including this zebra dataset) has found that when representing movement as static GPS points in statistical models we incorrectly capture how we infer animals use linear features (Holloway 2020). We could subsequently use this visualization technique to navigate other landscape features, such as roads, artificial water sources (i.e., wells), and rivers to study movement patterns around these locations, supporting research that is currently being undertaken in the area (Bennitt et al. 2021).

Dynamic visualizations within GIS will continue to advance and become more prevalent across fields. Visualization within a movement context relies heavily on the use of cartography, and in many respects the 'dynamic' video we have created is simply a series of 'snapshots' as outlined in the opening paragraph of this chapter. However, the finer resolution and dynamic nature of the spatiotemporal data allows us to explore this problem set in new ways. Representing movement as a static entity does not reliably account for the complex dynamic relationships that exist between movement and the environment, which means that conservation efforts developed off static maps for SDG15 may miss or overlook important details regarding these interactions. Temporal slider could also be used to represent aggregated data, such as country- or grid-level data, used in Chapters 8 and 9, if we had a temporal element to the attribute table. As such, the potential of the Temporal Controller to provide new insights across datasets is a very useful GIS skill.

10.3.1 Test Yourself

If you want to test yourself on the learning outcomes of this chapter, complete the following:

a. Generate another video focusing on a movement interaction with the landscape (i.e., roads, rivers, land cover) or other individuals (i.e., herding).

REFERENCES

Bartlam-Brooks, H.L., Beck, P.S., Bohrer, G. and Harris, S., 2013. In search of greener pastures: Using satellite images to predict the effects of environmental change on zebra migration. *Journal of Geophysical Research: Biogeosciences*, 118(4), pp. 1427–1437.

Bartlam-Brooks, H.L.A. and Harris, S., 2013. In search of greener pastures: Using satellite images to predict the effects of environmental change on zebra migration. Movebank Data Repository. doi: 10.5441/001/1.f3550b4f.

Bennitt, E., Bradley, J., Bartlam-Brooks, H.L., Hubel, T.Y. and Wilson, A.M., 2022. Effects of artificial water provision on migratory blue wildebeest and zebra in the Makgadikgadi Pans ecosystem, Botswana. *Biological Conservation*, 268, p. 109502.

Dodge, S., 2016. From observation to prediction: The trajectory of movement research in GIScience. *Advancing Geographic Information Science: The Past and Next Twenty Years*, pp. 123–136.

Holloway, P., 2020. Aggregating the conceptualization of movement data better captures real world and simulated animal–environment relationships. *International Journal of Geographical Information Science*, 34(8), pp. 1585–1606.

Holloway, P. and Miller, J.A., 2018. Analysis and modeling of movement. In *Comprehensive Geographic Information Systems*, Huang, B., Cova, T.J., and Tsou, M.H. (Eds.) Elsevier, Oxford.

Laube, P., 2014. *Computational Movement Analysis*. Springer International Publishing, Cham.

Section IV

Spatial Analysis: Measurements

11 Neighborhoods

11.1 INTRODUCTION AND LEARNING OUTCOMES

Neighborhoods are a central consideration for data analysis in GIS. We have already explored several analytical features related to the concept of a neighborhood in this book. For example, in Chapter 5, we identified all schools that were within a specific distance of a low broadband uptake area and in Chapter 7 we used cartography to present visualizations of litter within a specified study area to improve map balance. When considering the neighborhood within GIS, there are two elements that must be considered: distance and adjacency.

Distance is the measurement from one location to another in map units, while adjacency in spatial terms represents all locations that are considered immediate neighbors. We have already performed this type of spatial analysis in our case studies so far, and if we think back to Chapter 4, we should remember that the choice of projected coordinate system will greatly influence our distance measurements. However, we can also get different measurements based on the spatial data model used, as well as the method of measurement for both distance and adjacency (see Sarkar 2019 and Holloway & Miller 2018 for a detailed description on distance and adjacency, respectively).

The predominant method of calculating distance in GIS is Euclidean distance, or in other terms the straight-line distance. This is based on Pythagorean geometry, where we calculate the distance between AB as the hypotenuse (Figure 11.1a). This method is implemented in both vector and raster datasets, although as we know from the spatial structure of a raster, a distance unit is captured for every location, and is often referred to as 'proximity' (Figure 11.1c). Alternatively, and predominantly used with raster layers, is the Manhattan distance, where the distance is the length along the raster cells sides between AB, meaning it is impossible to pass diagonally (Figure 11.1b). This is perhaps best thought of using the gridded system of Manhattan, New York, where one must walk along the road network to get from A to B. Finally, given that the road network is not always gridded, in GIS it is possible to capture the network distance for vector datasets, particularly if movement is restricted over a network, such as a road or river system; we will explore this in Chapter 18.

Adjacency can be also represented in various manners, with the most common methods defining the neighborhood as the four immediate neighbors (von Neumann or rook's – Figure 11.1d) or the eight surrounding neighbors (Moore or Queen's – Figure 11.1e). Alternatively, the neighborhood can be defined as a specified distance buffer of various shapes that is representative of the movement capacities of the object, such as circle buffer (Figure 11.1f), which we briefly used in Chapter 5. Subsequently, with so many options to define neighbors, the choice of neighborhood definition can strongly influence the outcomes of any spatial analysis undertaken.

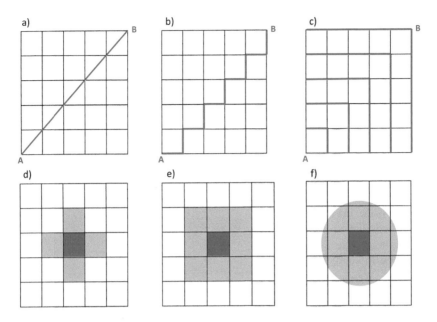

FIGURE 11.1 Conceptualizations of distance and adjacency used in GIS. (a) Euclidean distance between A and B, (b) Manhattan distance between A and B, (c) Proximity between A and B, (d) four immediate neighbors, (e) eight immediate neighbors, (f) circle buffer of neighbors defined by a distance.

Moreover, some of the adjacency conceptualizations, such as buffers, are built upon distance measurements, and therefore straddle both.

In this chapter, we work toward learning outcomes related to implementing neighborhood definitions using both vector and raster data, through the prism of SDG15.8. This target aims to introduce measures to prevent and significantly reduce the impact of invasive alien species. In addressing this, we use neighborhood definitions to highlight potential high-risk locations of invasive spread. By the end of this chapter, you will have completed four learning outcomes, and you should be able to:

- Digitize location information to create new data
- Generate raster data through the creation of grid and spatial queries
- Perform GIS operations related to neighborhood definitions, using both vector and raster conceptualization
- Use geometric operations to combine multiple layers

11.2 CASE STUDY: SDG15.8 IDENTIFYING POTENTIAL INVASION SITES THROUGH NEIGHBORHOOD ANALYSIS

The Argentine ant (*Linepithema humile*) is a highly invasive invertebrate species, with characteristics that provide it with a dominance over native species, threatening many ecosystems. In Spain, this invasive species has a largely coastal distribution due to ambient climate, but recent research has identified that it has spread to many

TABLE 11.1

Location of Surveyed Sites with Confirmed Sightings of the Argentine ant from López-Collar and Cabrero-Sañudo (2021)

Site	Latitude	Longitude	Source
1	40.417	−3.734	S
2	40.4189	−3.7223	S
3	40.411	−3.691	S
4	40.48072	−3.69557	S
5	40.38853	−3.68268	S
6	40.43533	−3.69233	S
7	40.452	−3.716	B
8	40.387	−3.759	P
9	40.411	−3.68888	S
10	40.42303	−3.68043	S
11	40.45933	−3.67263	S
12	40.037	−3.605	B
13	40.45833	−3.80467	P
14	40.359	−3.547	P
15	40.445	−3.998	P

Source documented in López-Collar and Cabrero-Sañudo (2021) as follows: S, survey data from their study, B, bibliographic references from Collingwood and Yarrow (1969), Martínez et al. (1997), and Ruiz Heras et al. (2011), and P, the species was reported to López-Collar and Cabrero-Sañudo (2021) via personal communication, according to N. Trotta leg. By J. Reyes, J. Arcos, K. Gómez and M. Sierra.

cities, where the urban heat island effect has been posited to negate the unfavorable climate conditions needed for its survival (López-Collar & Cabrero-Sañudo 2021). The recent Intergovernmental Science-Policy Platform on Biodiversity and Ecosystem Services (IPBES) report (Diaz 2019) identified that urban areas are one of the most impacted landscapes for biodiversity, meaning efforts are needed to stall the spread of invasive species. In their recent paper, López-Collar and Cabrero-Sañudo (2021) identified 15 locations where the Argentine ant has been surveyed in Madrid, Spain.

In this chapter, there is no data provided, in part due to sharing permissions associated with the urban atlas that we use later. By providing the location of ants in text format, it offers an opportunity to create new data through the process of digitization.

1. Open QGIS and start a New Empty Project.
2. Save the project as Ch11.

The first thing we need to do is install a new plugin called NumericalDigitize.

3. Navigate through the tab Plugins > Manage and Install Plugins...
4. Search for NumericalDigitize, and install it.

Next, we need to ensure the 'Digitize' toolbar is activated.

5. Right click on the gray area (see Chapter 2 for a refresher if needed) to open a new toolbar.
6. Make sure that 'Digitizing Toolbar' is ticked.

Next, we need to create a new vector layer that will represent the ant locations.

7. Navigate through the tab Layer > Create Layer > New GeoPackage Layer.
8. Create a new GeoPackage called Ch11_Data and save the layer as 'AntPoints'.
9. Set Geometry Type to Point.
10. Set the EPSG code to 4326 (WGS84).

Both should be the default, but it's worth double checking. At this point, we do not need to add any fields, we are only interested in the latitude and longitude values. We should note the warning message that states that this projection has somewhat limited accuracy. For now, we can ignore this.

11. Ensure your dialogue box resembles Figure 11.2 and click OK.

We should see the new layer in the Layers Panel, as shown in Figure 11.3.

12. Click on the layer so that it is selected in the Panel and highlighted blue.
13. Next, click Toggle Editing in the Digitizing Toolbar.

This should activate several of the tools. We are interested in the Numerical digitize tool.

14. Click on the Numerical digitize tool.

A dialogue box should open, which is where we add the coordinates of the points.

15. Add in the coordinates for site 1 from Table 11.1. Remember that latitude is Y and longitude is X.
16. Ensure your dialogue box resembles Figure 11.4. Click OK.

The dialogue box shown in Figure 11.5 opens. Because we did not specify any attributes when we created the layer, we cannot add any information to it. At this point in our GIS journeys, this is not an issue as we are not collecting or storing any further information on the point locations.

Neighborhoods

FIGURE 11.2 Screenshot of the parameters to add a new geopackage and layer.

FIGURE 11.3 Layers panel that contains the newly created layer.

17. Click OK.

A single point should have been created in the center of the map. We might want to give this context.

18. Turn on the OSM basemap from XYZ Tiles and zoom into the point. We may need to reorder the layers in the Layers Panel.

The first point should be situated in the park near the 'Lago de la Casa de Campo' in Madrid, Spain, as shown in Figure 11.6. If not, check that the coordinates have

FIGURE 11.4 Screenshot of the first point coordinates in the numerical digitize tool.

FIGURE 11.5 Screenshot of the autogenerate option for the unique ID (FID).

FIGURE 11.6 Screenshot of the QGIS map canvas with the first point located near the Lago de la Casa de Campo. Basemap is the OpenStreetMap XYZ tiles which is © OpenStreetMap contributors and available under the Open Database License. Please see https://www.openstreetmap.org/copyright.

Neighborhoods

been added in the correct format (X = longitude, Y = latitude). The symbology of the points has also been changed for clarity purposes to red and size 5.

19. Add the remaining 14 locations using this method. This must be completed one point at a time. There is an Add Row button within the dialogue box, but this does not always update all points.
20. Once this is completed, save the edits using the Current Edits button in the Digitizing Toolbar.

Once all points have been added, you should have a data layer that resembles Figure 11.7. If you have some points that are not in Spain, check you have added the '-' to the longitude values.

An alternative method would be to save Table 11.1 as a *.csv file and add the layer as a delimited text file, as we have undertaken in previous chapters. However, we input the data using this method to demonstrate further GIS functionality, as well as to highlight that there are often multiple avenues to reach the same result in GIS. Digitization is also a skill that is frequently required when we do not have existing spatial data. There are several tools in the digitization toolbar that can be used without a specific plugin, but require us to draw freehand, inferring that we know the location of the features we are digitizing.

Argentine ants disperse through local movements and long-distance human-mediated events, such as in accidental soil transportation. One solution to managing this invasive species is to monitor locations within suitable neighborhoods as to their susceptibility of invasion. The local dispersal capacity is approximately 150 m per year, while the long-distance dispersal events can range to several hundred

FIGURE 11.7 Screenshot of all 15-point locations digitized in Madrid. Basemap is the OpenStreetMap XYZ tiles which is © OpenStreetMap contributors and available under the Open Database License. Please see https://www.openstreetmap.org/copyright.

kilometers depending on the transportation method (Suarez et al. 2001). With policy recommendations for sustainable cities, the idea of the 15-minute city has become prominent in urban planning (Moreno et al. 2021). This concept advocates for an urban set-up where locals have access to most basic amenities within a 15-minute walk or cycle. If we consider distance as a function of the 15-minute city, the average distance covered by walking in 15 minutes is approximately 1.2 km, while the average distance covered by road (which could be car or bike) using the 50 km/hour speed limit of Madrid is 12.5 km. While a 15-minute car journey is not technically within the confines of the sustainable city concept, it is an appropriate distance one might drive to access green space, and therefore considered in this analysis.

Therefore, to assess the neighborhoods surrounding the existing ant locations, we create three distance buffers of 150, 1200, and 12,500 m. However, and hopefully you have realized this already, our data is unprojected and currently only recorded in WGS84. Therefore, we first must project our data to a suitable Cartesian format. We are going to stick with UTM as we have explored throughout the book, so we need to project our data to UTM Zone 30N.

21. Navigate through the tab Vector > Data Management Tools > Reproject Layer.
22. Set AntPoints as the Input layer.
23. Change the Target CRS to EPSG 32630.
24. Save the output as AntPointsUTM to the GeoPackage.
25. Ensure the dialogue box resembles Figure 11.8. Click Run.
26. Change the projection of the map canvas as well in the bottom right-hand corner to UTM Zone 30N.

Next, we process our buffers. For visualization purposes, the points visualized herein have been symbolized as blue and size 5 to differentiate them from the unprojected layer presented earlier.

27. Navigate through the tab Vector > Geoprocessing Tools > Buffer.
28. Specify AntPointsUTM as the input layer.
29. Specify 150 m as the distance.
30. Make sure Dissolve Output is ticked.
31. Save the output layer in the GeoPackage with a name that indicates the buffer distance, such as AntBuffer150m.
32. Ensure your dialogue box resembles Figure 11.9. Click Run.
33. Zoom to the buffer.

A 150 m buffer has been placed around the location of the ants, with this data now represented as a polygon file. This layer represents the area within 150 m of each of the surveys.

34. Repeat the buffer analysis for 1200 and 12,500 m using steps 27–32 (but obviously updating the name of the variables to represent the new distances).

Neighborhoods

FIGURE 11.8 Screenshot of parameters for the reproject layer tool.

FIGURE 11.9 Screenshot of parameters for the buffer tool for the 150 m buffer.

35. Rearrange the order of the variables in the Layers Panel so that 150m is at the top and 12,500m is at the bottom.

Your output should resemble Figure 11.10. In cartographic terms, this type of visualization is an isochrone, which is a type of map that depicts the area accessible from a specific point within a specified timeframe.

When specifying the parameters of the buffer tool, we ensured that 'dissolve result' was ticked. This option dissolves boundaries that have similar values, and because this is a buffer, we do not need to retain any of the information previously stored in the point layer. If we repeat the buffer process for 12,500m, but do not dissolve the boundaries, we can see the difference in results in Figure 11.11.

This increases the amount of information held within the GIS, and ultimately creates 15 different polygons instead of two to represent the same information. It is also quite messy. Therefore, the dissolve option has two benefits, it presents our work clearly and removes data redundancy. This latter point is important, not necessarily in this example, but in examples where we may be working with thousands of points, lines, or polygons. In these instances, not dissolving our buffer

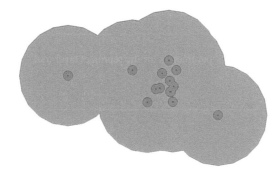

FIGURE 11.10 Screenshot of the three buffer distances for northern Madrid.

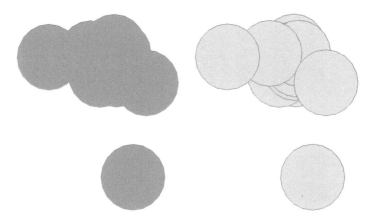

FIGURE 11.11 Comparison of 12,500m buffer with and without dissolve applied.

can introduce a large amount of redundancy, and unnecessarily slow down subsequent analysis. This is a common and easy mistake that is often made in a lot of my classes and can easily add 20+ minutes to processing times – a simple omission to not ensure that the results are dissolved when buffering can cause real headaches later down the analysis pipeline. This becomes more evident in later chapters when we use buffers as part of a wider analytical framework. Therefore, unless you have a good reason not to dissolve buffers, I suggest implementing this every time.

Now we have our 'neighborhoods' defined; we can explore what locations are found within them. Obviously, the larger the buffer, the more possible locations are returned. Given the influence that the Argentine ant has on native insects, we might be interested in locations within our buffers/neighborhoods that are of a certain land cover. Green areas are vitally important to urban biodiversity, meaning one way of supporting SDG15.8 would be to quantify the number of possible urban green areas that are 'at risk' of invasion based on the three neighborhoods.

If we turn our basemap layer back on, we can visually inspect these areas. However, we can also use GIS to generate new data that is only found within the buffers. To do this, we use the European Environment Agency (EEA) Urban Atlas. The Urban Atlas is a high-resolution land cover and land use dataset for 788 European urban areas that has been made available for 2006, 2012, and 2018 (EEA 2021). To access this data, we first need to set up an EEA account, and then download the data for Madrid.

36. Navigate to https://land.copernicus.eu/local/urban-atlas/urban-atlas-2018?tab=mapview.

This takes us to a map view where we can navigate the different cities which have data available.

37. Click the Download Tab.
38. This webpage will ask you to log in to download the data. If you have an EEA account, please login, else register using your details.
39. Upon completion, you should be returned to the following page shown in Figure 11.12.
40. Search for Madrid.
41. Tick the box to the left of the city.
42. You will also be asked to tick another box stating that you are aware you are downloading non-validated data, and that you accept not to distribute the information contained within these files or the files themselves further. This is a legal agreement, so by checking this, you are agreeing to the terms and conditions of the EEA. If you are not comfortable agreeing to this, then that is not a problem, but please proceed to the next chapter.
43. Once both boxes are checked, select 'Download'.

Once complete, this will automatically download. Depending on the internet connection and speed, this could take a few seconds or minutes.

FIGURE 11.12 Download page for the Urban Atlas 2018 datasets.

44. Once the data is downloaded, navigate to the folder, and extract it.
45. Upon extraction, open the folder. There is another zipped folder here named 'ES001L3_MADRID_UA2018_v013'. Extract this folder.

Within this folder there are four sub-folders: Data, Documents, Legend, and Metadata. Familiarize yourself with the data should you wish. We are going to add the data to QGIS.

46. Navigate through the tab Layer > Add Layer > Add Vector Layer. Navigate to Data, and select *.gpkg.

There are three layers saved here. We want to add the ES001L3_MADRID_UA2018 layer, the one with 105,392 features.

47. Select the ES001L3_MADRID_UA2018 layer.
48. Click OK.
49. A transformation box will also open. We can select the default transformation for the projection by clicking OK.

Your map should resemble Figure 11.13.

Within the downloaded files, there is a sub-folder called 'Legend'. This is a predesigned cartographic symbology for the different layers. In Chapter 9 we created and saved a symbology file in case we needed to reapply it. Here, the EEA have created a formal symbology to represent the urban land cover categories, ensuring consistency in visualization across all 788 cities. To improve visualization, we should apply it.

50. Right click on the layer in the Layers Panel, and select Properties.
51. Select the Symbology tab.

Neighborhoods

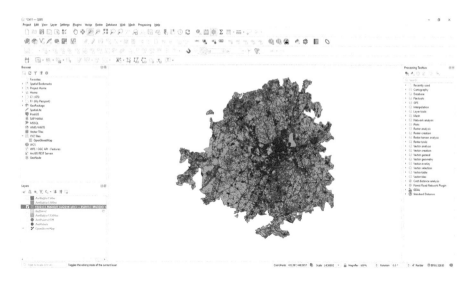

FIGURE 11.13   Screenshot of QGIS project with the 2018 Urban Atlas for Madrid loaded.

52. Change the style from Single Symbol to Categorized.
53. Select Value as 'code_2018'.
54. Click Classify.

This loads several different codes, each one representing a land cover class. Next, we upload the layout color palette.

55. Click Style > Load Style.
56. Click on File …
57. Navigate to where the folder was extracted to. In the sub-folder Legend, select the Urban_Atlas_2018_Legend file.
58. Click Load Style.

The codes have been updated with the predetermined legend and color palette. If we look at the different categories as shown in Figure 11.14, we can see that there are in fact several land cover classes that might be of interest to this investigation. For the moment we focus on Green urban areas (code 14100).

59. Click Apply to apply the color palette.

Your map should now resemble Figure 11.15. This symbolization is much clearer, and we can immediately see the 'greener' areas of the city, particularly in the suburban areas. This is a further example of how cartography can support spatial cognition. It is also a real-world application of utilizing a pre-existing color palette that we explored in Chapter 9.

We want to identify all green areas that are within our Argentine ant neighborhoods. Firstly, we must select all the features that match the criteria of 'Urban green

190 Understanding GIS through Sustainable Development Goals

FIGURE 11.14 Screenshot of loaded Urban_Atlas_2018_Legend file in the symbology tab of properties.

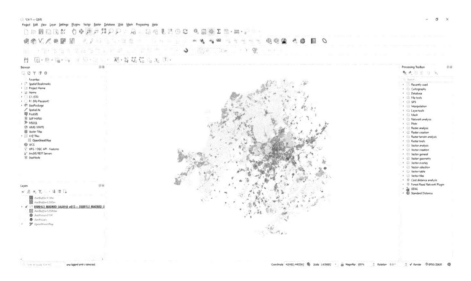

FIGURE 11.15 Screenshot of the Urban Atlas 2018 with loaded legend file applied.

Neighborhoods 191

area' and export it as a new layer. We should hopefully be familiar with the subsequent steps by now if we are working chronologically through the book, so try and select all the features that match 'urban green areas' without looking at steps 60–66 but noting that the Code_2018 for this land cover is 14100.

60. Highlight ES001L3_Madrid_UA2018 in the Layers Panel.
61. Click the Select by Values button.
62. Type 14100 in Code_2018.
63. Change the operator to 'Equal to'.
64. Click Select Features.
65. Right click on the layer in the panel, Extract > Save selected features as.
66. Save this in the GeoPackage created earlier, with the layer name UrbanGA.

To check that this has worked correctly, zoom into some of the areas to visually match up the new UrbanGA layer with existing green areas in the city-wide urban atlas.

67. Turn off ES001L3_MADRID_UA2018 to speed up processing.

Finally, we want to ensure that the UrbanGA layer is in the same projection as AntsUTM. Therefore, we need to project this to UTM 30N. Again, try to complete the following four steps independently if possible.

68. Navigate through the tab Vector > Data Management Tools > Reproject Layer.
69. Select UrbanGA as the Input Layer.
70. Set the Target CRS as 32630.
71. Save the reprojected layer in the GeoPackage as UrbanGA_UTM.

Next, to identify green areas at possible risk of invasion, we quantify the amount of this land cover within the three neighborhoods. Again, we start the instructions with the natural spread model of 150 m. To perform this operation, we use a geoprocessing tool called clip. When we clip a layer, we create a new layer that has a new extent that conforms to the boundaries of another layer. Clip uses that other layer as a 'cookie cutter' and is useful for trimming a layer down to isolate a particular area of interest. Clip results in an output layer that has the same feature type as the input layer (i.e., polygon to polygon, points to points). The input layer can consist of point, line, or polygon features, but the clipping layer must be a polygon.

72. Navigate through the tab Vector > Geoprocessing > Clip.
73. Select the UrbanGA_UTM as the input layer.
74. Select AntBuffer150m as the overlay layer.
75. Save the clipped layer to the GeoPackage, again specifying the distance in the name, such as UrbanGA150m.
76. Ensure the dialogue box resembles Figure 11.16. Click Run.

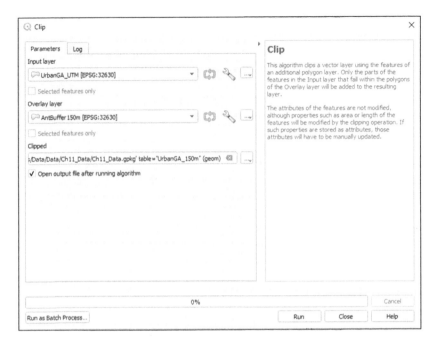

FIGURE 11.16 Screenshot of parameters needed for the clip tool with the 150 m neighborhood.

Upon completion, only a small subset of the UrbanGA remains. This is the cookie cutter process, whereby only locations within the 150 m buffer overlay layer are returned. To see how many separate green areas are within this distance, open the attribute table.

77. Right click on UrbanGA150m in the Layers Panel, and open the attribute table.

There are 25 features within the table. While some of the 15 ant locations were found in green spaces, upon completion of this analysis we now know that there are at least 25 green areas that could be considered 'at risk' of invasion through the natural spread of the species.

78. Repeat the clip for the other two neighborhood sizes, again saving the clipped layer with an informative name.

For reference, there are 328 features returned when considering 1.2 km (or 1200 m) and 2659 features when considering 12.5 km (12,500 m). If UrbanGA has not been reprojected, there are 2841 features for 12.5 km, again highlighting the importance of a consistent projection.

Here we begin to see the difficulty in managing many invasive species, especially once they have become established in an area. We have successfully used GIS to quantify the locations of established Argentine ant populations in Madrid and the location of nearby green areas where populations have not yet spread but are within the dispersal capacities either naturally or through human-mediated movement. As such, the locations within 150 m of established populations could be the focus of regular monitoring to ensure that the natural spread of the species is minimized and support eradication efforts, which is common practice (Boser et al. 2014). Similarly, the locations within 15 minutes either by foot or road could be subject to an information drive, whereby signage and information is posted asking for recreational users to check their boots, wheels, and other material that might contain soil and/or other material where eggs or ants could be unintentionally transported. Such approaches have been successful in managing other invasive species, such as the zebra mussel (Carey 2009). To specifically quantify this through a target, signage could ask people to report any sightings, which could lead to a successful number of spread events that have been thwarted by such measures.

Thus far, this chapter has focused on how to calculate distance buffers on vector data. The remainder of the chapter will replicate this analysis using raster data. The first thing we need to do is convert our point data to raster format. Again, there are various ways of achieving this. We could use the tool 'Rasterization' directly on the point data, as we explored in Chapter 3, but to demonstrate different ways of using GIS we work through an alternative method. If we have not done so already, change the EPSG code in the bottom right-hand corner of the map canvas to UTM 30N (EPSG: 32630).

79. In the Processing Toolbox navigate to Vector creation > Create grid.
80. Double click on the Create grid tool.
81. Select Grid Type as 'Rectangle (polygon)'.
82. Click on the Grid Extent options ... and Select 'Calculate from Layer'. Here choose the layer that covers the full extent of Madrid, ES001L3_MADRID_UA2018.
83. Specify the horizontal and vertical spacing as 1000 m.
84. Ensure the output CRS is UTM zone 30N.
85. Save the output file to the GeoPackage as Grid.
86. Ensure your dialogue box resembles Figure 11.17. Click Run.

A regular lattice grid appears over the extent of Madrid, with each rectangle 1000 m in width and height. We could of course increase or decrease the resolution of this grid (i.e., make the size larger or smaller), but for this case study 1 km is sufficient. Next, we populate this grid with information regarding the presence of the Argentine ant.

87. In the Processing Toolbox or Selection Toolbar open the Select by location tool.
88. Select Grid from the Select Features from options.
89. Select the geometric predicate as 'intersect'.
90. Select the 'comparing features from' as AntPointsUTM.
91. Click Run.

A small number of grids are now selected. We want to generate a new value for these fields only.

FIGURE 11.17 Screenshot of parameters for the create grid tool.

92. Right click on the Grid layer in the Layers Panel, and open the attribute table.
93. Open the field calculator.
94. Ensure that 'Only update 15 selected features' is ticked.
95. Tick 'Create a new field'.
96. Name the Output Field name as Ants.
97. And type 1 in the Expression.
98. Ensure your dialogue box resembles Figure 11.18. Click OK.

In the attribute table, there should be a new column called 'Ants', with all the selected features having a value of 1, with all the other rows reporting 'NULL'. We now want to change the null values to a numerical value. We can do this quite simply by inverting the selection and repeating these steps, but this time we type 0 in the expression.

99. Click the Invert Selection button.
100. Open the field calculator.
101. Ensure that the 'Only update 17932 selected features' is ticked.
102. Tick 'Update Existing Field'. We have already created the Ants attribute, so we do not need to do it again. Select Ants as the field to update.
103. Type 0 in the Expression.
104. Your dialogue box should now resemble Figure 11.19. Click OK.

We now have a complete attribute table of 0 or 1s. This is a binary representation of presence (1) or absence (0) of ants. Next, we need to convert this to a raster.

Neighborhoods

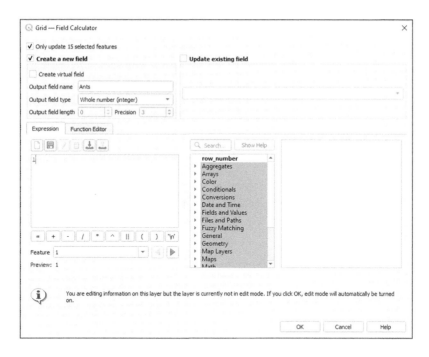

FIGURE 11.18 Screenshot of parameters for the field calculator.

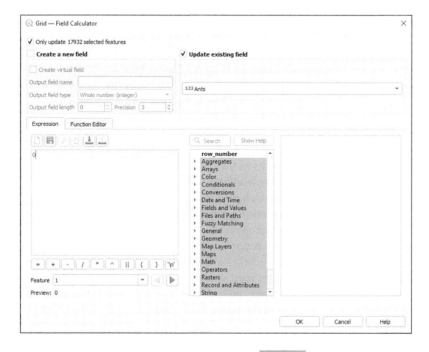

FIGURE 11.19 Screenshot of parameters for the field calculator.

FIGURE 11.20 Screenshot of parameters for the rasterize tool.

105. Click the Deselect All Features from All Layers button.
106. Save the edits in the Digitizing Toolbar, otherwise the analysis will not work. We do this by deselecting Toggle Editing, and selecting save.
107. Navigate through the tab Raster > Conversion > Rasterize.
108. Choose Grid as the input layer.
109. Select the newly created Ants field to use for a burn-in value.
110. Choose georeferenced units as the output raster size.
111. Set the width and height resolution to 1000 m.
112. Set the output extent to match grid.
113. Save the output in your working directory as antRaster. Remember, we cannot save rasters in GeoPackages.
114. Ensure your dialogue box resembles Figure 11.20. Click Run.

This should have created a raster representation of the data. The burn-in value is the only value displayed in our raster; however, we can still use this layer to calculate the distance at every location in the raster from one of the cells with an ant presence confirmed.

115. Navigate through the tab Raster > Analysis > Proximity.

Neighborhoods

116. Set antRaster as the Input Layer.
117. Set Distance Units to Georeferenced coordinates.
118. Save the output layer in your working directory, as antDistance.
119. Keep the rest of the options as default.
120. Ensure your dialogue box resembles Figure 11.21. Click Run.

Upon completion, we have a continuous raster representation of the distance from confirmed ant locations, as shown in Figure 11.22. Consult the Layers Panel to see that the distance is calculated on a grayscale, with smaller distances represented as darker colors, with the color getting lighter as the distance increases. We can use the Identify Features tool to explore the values.

Using the same investigative research question to identify cells that fall within the specific neighborhoods, we use the reclassify tool. This tool changes the raster layer values to new values based on a set of specified criteria, such as aggregating values into distance bins.

121. In the Processing Toolbox, search for Reclassify by table.
122. Open the tool.
123. Set antDistance as the input layer.
124. Choose options for the Reclassification table ... This should open a new dialogue box.

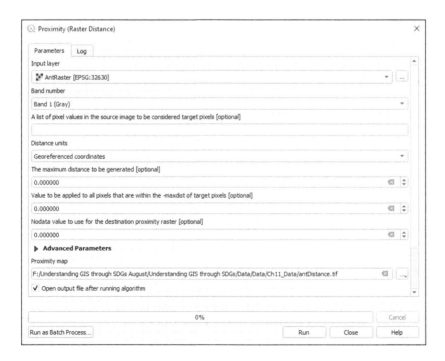

FIGURE 11.21   Screenshot of parameters for the proximity tool.

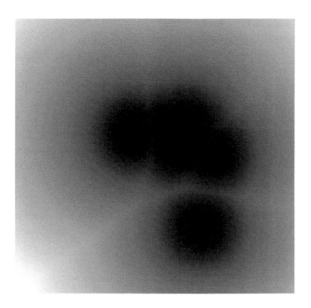

FIGURE 11.22 Screenshot of QGIS upon completion of the proximity tool.

Here we set the minimum and maximum of the different bands we wish to reclassify based on the three neighborhood sizes, as well as Value, which will be the 'new value' we create. We must add 1 to each of the neighborhood sizes to ensure that the smallest neighborhood is classified.

125. Complete the table as outlined in Figure 11.23.
126. Click OK.

Clicking OK is a key component as otherwise it will not save the table and the tool will fail. In other words, do not run the tool with this view.

127. In Advanced Parameters, ensure that Range Boundaries is selected as: min <= value < max.

This will ensure that the ranges do not overlap each other and removes the ambiguity that we came across in Section 3 when dealing with values straddling boundaries.

128. Tick 'use no data when no range matches value'. This will revert a NoData value for any values that are greater than 12,500 m.
129. Save the output in the working directory as AntReclass.
130. Ensure your output resembles Figure 11.24. Click Run.

We now have a new raster layer with the largest value symbolized as white. To aid spatial cognition, we may want to change the symbology to make this more visual.

Neighborhoods

FIGURE 11.23 Screenshot of Reclassify by table parameters.

FIGURE 11.24 Screenshot of parameters needed for reclassify by table tool.

131. Right click on the AntReclass layer, and open properties.
132. Navigate to the Symbology tab.
133. Select the Render Type as 'Palleted/Unique values'.
134. Change the color palette to Spectral.
135. Click Classify.
136. This will show the three unique values.
137. Click OK.

Your new layer should now resemble Figure 11.25.

The final step in this chapter is to clip this raster to the urban green areas, so that we can see the amount of area within those neighborhoods.

138. Navigate through the tab Raster > Extraction > Clip Raster by Mask Layer.
139. Select the AntReclass layer as the input.
140. Select the UrbanGA_UTM as the mask layer.
141. Set the Source and Target CRS as UTM Zone 30N.
142. Save the output to your working directory as AntReclass_Clip.
143. Click run.

Again, the output will be generated on grayscale, so to visualize this new layer intuitively we should change the symbology to Spectral again.

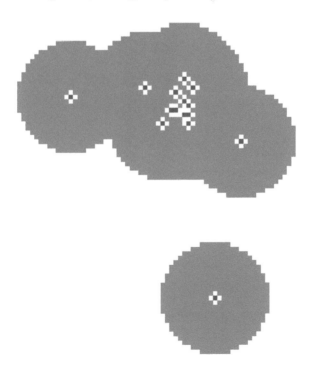

FIGURE 11.25 Screenshot of the AntReclass layer symbolized using the spectral color palette.

Neighborhoods

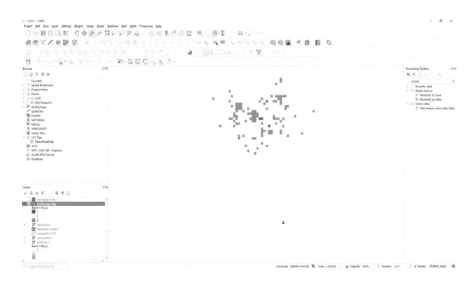

FIGURE 11.26 Screenshot of AntReclass_Clip layer.

144. Repeat steps 131–137 for AntReclass_Clip.

Your results should resemble Figure 11.26. There is not an attribute table associated with raster data structures, so we cannot simply identify the amount of green area features selected. Instead, we use another tool.

145. In the Processing Toolbox navigate to Raster Analysis > Raster layer unique values report.
146. Specify AntReclass_Clip as the Input Layer.
147. Keep the outputs as temporary layers.
148. Click Run.

The Results Viewer should appear with a html link.

149. Open the html link.

From this output, we can see the results as reported in Table 11.2. Here we can see that there are 5 pixels of green space that are within the 150 m neighborhood, 9 pixels within 1200 m, and 101 within 12,500 m. This function has also calculated the area of 5 million, 9 million, and 101 million m^2. Remember, due to the reclassification scheme, the larger neighborhoods actually have more pixels in them; i.e., the 1200 m also contains the 5 pixels that are specified within the 150 m neighborhood. Due to the resolution of the raster being quite coarse (1000 m) compared to the urban green areas, it does appear that this quantification may have overestimated the amount of area. This is quite common when working with raster data as it assumes homogeneity within each raster grid cell, which is not always the case. When we consider this along with the visual output of the AntReclass_Clip, it also seems as if this

TABLE 11.2
Results from the Raster Layer Unique Values Report

Value	Pixel Count	Area (m²)
1	5	5,000,000
2	9	9,000,000
3	101	101,000,000

geoprocessing has not fully captured all the relevant information. This is an important consideration and reinforces the significance of selecting an appropriate spatial data model to undertake our analysis. Moreover, we should always consider the scale at which our data is generated at. The raster approach may have been more informative if we had used a 10 m or 100 m resolution as opposed to a 1000 m resolution.

11.3 CASE STUDY CONCLUDING REMARKS

Obviously, as the size of the neighborhood increases, the amount of green area that could be 'at risk' also increases. While not surprising, it does provide city managers with a simple tool to assess areas that may need regular monitoring and management. Once invasive species have established themselves in a region, limiting their spread is one of the best mitigation efforts, and by using buffer and proximity tools to identify areas within a natural dispersal distance and 15 minutes of human-mediated transportation, we have identified areas within a realistic distance whereby dispersal is more likely to occur. Moreover, we have calculated this for green areas, a landscape that is increasingly important for native biodiversity in urban areas (Díaz et al. 2019; Lambert et al. 2021).

This chapter has reinforced and introduced methods to generate data through digitization and grid creation, as well as reinforcing the buffer tool and introducing us to the geoprocessing tool clip. We have undertaken analysis using both vector and raster data and highlighted the importance of scale in both spatial data models when considering neighborhoods.

11.3.1 Test Yourself

If you want to test yourself of the learning outcomes of this chapter, complete the following:

a. Repeat the raster analysis in steps 79 onward, but this time generate a smaller resolution (e.g., 100 m). Upon completion, you can observe how the results change. Please note, however, that this will require more computing power than was needed for a 1000 m grid, especially if you use a very fine resolution (e.g., 10 m).

b. Are there any other green areas in the Urban Atlas that should be incorporated in the analysis? Write a selection query to return multiple land covers and see whether this changes the results using the vector approach.

REFERENCES

Boser, C.L., Hanna, C., Faulkner, K.R., Cory, C., Randall, J.M. and Morrison, S.A., 2014. Argentine ant management in conservation areas: Results of a pilot study. *Monographs of the Western North American Naturalist*, 7(1), pp. 518–530.

Carey, J., 2009. *Stopping the Spread: Examining the Effectiveness of Policies in the Mississippi River Basin Aimed at Preventing the Spread of Zebra Mussels*. Oklahoma State University.

Collingwood, C.A. and Yarrow, L.L., 1969. A survey of Iberian Formicidae (Hymenoptera). *EOS. Revista Española de Entomología*, 44, pp. 53–101.

Díaz, S., Settele, J., Brondízio, E.S., Ngo, H.T., Guèze, M., Agard, J., Arneth, A., Balvanera, P., Brauman, K., Butchart, S.H. and Chan, K.M., 2019. Summary for policymakers of the global assessment report on biodiversity and ecosystem services of the Intergovernmental Science-Policy Platform on Biodiversity and Ecosystem Services. IPBES Secretariat, Bonn, Germany. https://ri.conicet.gov.ar/handle/11336/116171

EEA (2021). *Urban Atlas LCLU 2018, Version v013*. Release Date: 16-07-2021. European Environment Agency. https://land.copernicus.eu/local/urban-atlas/urban-atlas-2018?tab=mapview.

Holloway, P. and Miller, J.A., 2018. Analysis and modeling of movement. In *Comprehensive Geographic Information Systems*, Huang, B., Cova, T.J. and Tsou, M.H. (Eds.), Elsevier, Oxford.

Lambert, L., Cawkwell, F. and Holloway, P., 2021. The importance of connected and interspersed urban green and blue space for biodiversity: A case study in Cork City, Ireland. *Geographies*, 1(3), pp. 217–237.

López-Collar, D. and Cabrero-Sañudo, F.J., 2021. Update on the invasion status of the Argentine ant, *Linepithema humile* (Mayr, 1868), in Madrid, a large city in the interior of the Iberian Peninsula. *Journal of Hymenoptera Research*, 85, p. 161.

Martínez, M.D., Ornosa, C. and Gamarra, P., 1997. Linepithema humile (Mayr 1868) (Hymenoptera: Formicidae) en las viviendas de Madrid. *Boletín de la Asociación Española de Entomología*, 21, pp. 275–276.

Moreno, C., Allam, Z., Chabaud, D., Gall, C. and Pratlong, F., 2021. Introducing the "15-Minute City": Sustainability, resilience and place identity in future post-pandemic cities. *Smart Cities*, 4(1), pp. 93–111.

Ruiz Heras, P., Martínez Ibáñez, M.D., Cabrero-Sañudo, F.J., and Vázquez Martínez, M.Á., 2011. Primeros datos de Formícidos (Hymenoptera, Formicidae) en parques urbanos de Madrid. *Boletín de la Asociación Española de Entomología*, 35(1–2), pp. 93–112.

Sarkar, D., 2019. *Distance Operations. The Geographic Information Science & Technology Body of Knowledge* (3rd Quarter 2019 Edition), Wilson, J.P. (Ed.). University Consortium for Geographic Information Science, Online. http://gistbok.ucgis.org/.

Suarez, A.V., Holway, D.A. and Case, T.J., 2001. Patterns of spread in biological invasions dominated by long-distance jump dispersal: Insights from Argentine ants. *Proceedings of the National Academy of Sciences*, 98(3), pp. 1095–1100.

12 Descriptive Statistics

12.1 INTRODUCTION AND LEARNING OUTCOMES

Descriptive statistics are particularly useful for analyzing spatial data, as they quantify patterns for both the spatial and attribute information. The premise of descriptive statistics is to describe and summarize the characteristics of the data available to us (often termed a sample). The most common way of describing a variable's distribution is in terms of its two properties: central tendency and dispersion. Measures of central tendency attempt to find the location of the middle point in a dataset, such as the mean, median, and mode, while measures of dispersion describe how the observations are distributed, such as the variance and standard deviation.

Most of us have hopefully had some experience working with these statistics before, but most likely in a non-spatial setting. See Rogerson (2020) for a detailed account of these statistics in geographic research, for example, calculating the average age, height, or grades of a sample. The concept is exactly the same with spatial data, but instead of calculating the statistics on the aspatial data (or attribute data), we calculate it on the spatial data (i.e., coordinates). In most GIS software, we can calculate three main types of descriptive statistics:

Mean Center – Here, we calculate the average (or mean) x value and the average (or mean) y value. These become our mean centers, or centroids. This is often useful when we want to compare different features spatially, across attributes, or identify directional trends over time. In QGIS, we can only calculate the mean, but the concept can easily be extended to median or mode.

Standard Distance – Here, a value representing the distance in map units from the mean center is generated and is similar to how standard deviation measures distribution around the mean. This function quantifies the degree to which features are dispersed around the mean center. Therefore, the standard distance generally represents one standard deviation (approximately 68% of data) around the mean, although these can be altered within the functionality of the tool.

Standard Deviational Ellipse – Here, a value representing the directional trend of the dispersion of the data is provided. In this instance, the standard distance is calculated on the x and y axes and calculates the standard deviation around the mean center to define the axes of the ellipse.

These analytical techniques are widely used in criminology and epidemiology (Lek-Uthai et al. 2010; McGrath et al. 2014), and in this chapter, we use these techniques to investigate patterns in traffic accidents. This work relates to SDG3 Good Health and Wellbeing, which aims to ensure healthy lives and promote wellbeing for all ages.

Target SDG3.6 aims to halve the number of global deaths and injuries from road traffic accidents. If we can identify patterns in the location of traffic accidents and investigate whether these patterns change according to time of day and/or severity, we can take focused action to support target SDG3.6.

By the end of this chapter, you will have completed three learning outcomes, and you should be able to:

- Perform spatial statistics, including mean center, standard distance, and standard deviational ellipse
- Use the attribute table to weight analytical processes by certain geographic features
- Deconstruct spatial analysis across attributes to reveal hidden patterns

12.2 CASE STUDY: SDG3.6 ANALYSIS OF THE SPATIAL PATTERN OF TRAFFIC ACCIDENTS IN DENVER, COLORADO, USA

1. Open a New Empty Project in QGIS.
2. Add the point layer TrafficAccidents from the Ch12_data GeoPackage.
3. Save the Project as Ch12.

Your QGIS project should resemble Figure 12.1. Most traffic accidents are reported to the administrative authority of the area, such as city, county council, or state. Many of these organizations have an open data policy, meaning there is a wealth of datasets that could be explored using these techniques. This chapter uses data from the City of Denver, Colorado, USA, with a shapefile of traffic accidents downloaded

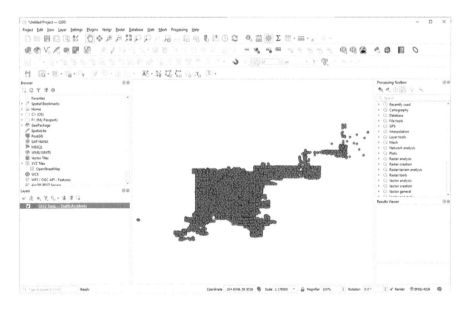

FIGURE 12.1   Screenshot of the TrafficAccidents layer imported into QGIS.

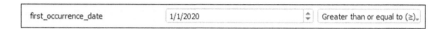

FIGURE 12.2 Screenshot of select features parameters to select all locations since 1 January 2020.

from the City of Denver Open Data Catalog under CC BY 3.0. The shapefile has been converted into a GeoPackage layer for use in this chapter.

This layer represents all traffic accidents in the City of Denver administrative area since 1 January 2013. In this dataset (downloaded in March 2022), since 2013 there have been approximately 209,773 traffic accidents, with 479 accidents resulting in 501 fatalities and 4386 accidents resulting in 5030 serious injuries. There is a wealth of information within this dataset, meaning we could explore trends in these statistics across the 6-year period. However, to make the analysis more manageable, as well as to focus our research question, we focus on accidents from 2020.

4. Open the attribute table to identify the date field (it is first_occurence_date).
5. Use the Select Features by Value button to open the dialogue box.
6. Specify an expression to return all values with an occurrence data greater than or equal to 1 January 2020, as shown in Figure 12.2.

If we click Select Features at this point, we return all the accidents since 1 January 2020, but this will also return accidents from subsequent years (i.e., 2021). In this dialogue box, there is no way to write another expression for the same attribute, meaning we need to find another way of achieving this.

7. Close the dialogue box, without clicking select.
8. Click on the Select Features by Expression button. If this is not visible, click on the downward triangle on Select Features to identify the other options.
9. Add the following expression, ensuring to use the quotation marks if not navigating to the fields and values directly in the middle panel: "first_occurrence_date" >= '2020-01-01' AND "first_occurrence_date" <= '2020-12-31'.

In this expression, we have used the AND operator to specify that we want to return all accidents that happen between 1 January and 31 December 2020. Ensure your dialogue box resembles Figure 12.3.

10. Click Select Features and then close the dialogue box.
11. Open the attribute table to see how many features have been selected.

There should be 15,929 selected features. If this is not the number of selected features, recheck the expression, and confirm that the correct operator has been used (i.e., >= instead of >).

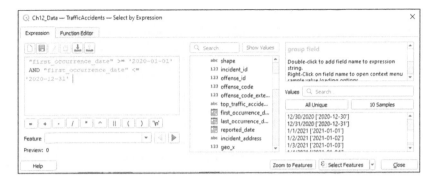

FIGURE 12.3 Screenshot of select features by expression parameters to select all locations in 2020.

12. Right click on the TrafficAccidents layer in the Layers Panel, and select Export Data > Save Selected Features As.
13. Save the new layer as Accidents2020 in the GeoPackage.

Now would be a good opportunity to add a basemap to provide some context for our accidents.

14. In the Browser Panel, turn on the XYZ tile for OpenStreetMap (the layers may need to be reordered in the Layers Panel).

We can begin to see the geographic context of where these accidents are occurring. Before we begin to analyze the data, we should note the artificial boundary of administrative areas. The administrative area of the City of Denver does not represent an island, meaning the boundary is somewhat artificially generated. There will most likely be accidents beyond the borders of this city. Therefore, any analysis we undertake on this dataset will be subject to the edge effects brought in by this boundary. There are various methods that can begin to account for such edge effects, but they are not always simplistic, and are often subject to data availability. For now, we continue our analysis, acknowledging that any analysis we undertake will be subject to missing data around the edge of the constituency.

To calculate the mean center of all the accidents in 2020:

15. Navigate through the tab Vector > Analysis Tools > Mean Coordinate(s).
16. Select Accidents2020 as the Input Layer.
17. For now, we can keep all the other features empty as default (including the temporary layer).
18. Ensure your dialogue box resembles Figure 12.4. Click Run.

This function should add a single point that represents the mean x and mean y coordinates. It is not obvious where this point is, so we should zoom to it.

Descriptive Statistics

FIGURE 12.4 Screenshot of parameters for the mean coordinate(s) tool.

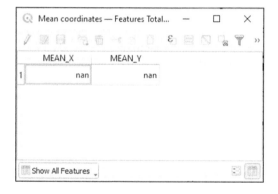

FIGURE 12.5 Screenshot of attribute table of mean coordinates layer.

19. Right Click on the mean coordinates layer, and click 'zoom to layer'.

Nothing happens... At this point, we should begin to think that our analysis has not worked. We should explore whether this is the case using the attribute table.

20. Open the attribute table of the mean coordinates layer.

The analysis has rendered a no data value, shown as 'nan' in Figure 12.5, meaning something is not quite right in our data. When this happens in our work, there are two components of the data we should check first: the projection and whether the original data contains null values. As this tool is independent of the projection, it is most likely null data. What we mean by this is that the tool is not dependent on two feature layers overlaying each other, so should run regardless of the projection. That is not to say the projection is not important, just that it should not impact processing of this tool.

21. Open the attribute table of Accidents2020.
22. Scroll across to Geo_Lat or Geo_Long.

The values in the Geo_Lat column should be around 39°, while in the Geo_Long column they should be around −104° to −105°.

23. Click Ascending (remember, this is simply clicking the column/attribute name).

There are almost 500 features that have a latitude and longitude of NULL. For the analysis to work, we need to remove these from our dataset. As we are removing NULL data, we also need to remove any features that have NULL for fatalities. When we reach the latter stages of this chapter, we use weights in the tools, but they will not work correctly if there are features with no data.

24. Close the attribute table.
25. Click on Select Features by Values by Expression.
26. Write the following expression: "geo_lon" < −100 AND "FATALITIES" >= 0.

This expression will return all features that have correct longitude and latitude and have reported data on the number of fatalities. We could choose −101, or even −104 in the expression, but this just gives us a bit of wiggle room.

27. Ensure your dialogue box resembles Figure 12.6. Click Select Features.
28. Right click on the Accidents2020 in the Layers Panel, Export > Save Selected Features As.
29. Save this layer as Accidents2020_Complete in the GeoPackage.
30. Repeat the steps to calculate the mean center, but this time using the layer Accidents2020_Complete as the input layer. We should also save the output in our GeoPackage with an informative name, MeanCenter_All.

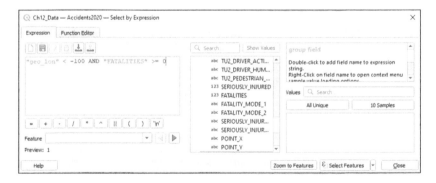

FIGURE 12.6 Screenshot of parameters needed for the select by expression to return values with valid coordinate data and data on fatalities.

Descriptive Statistics 211

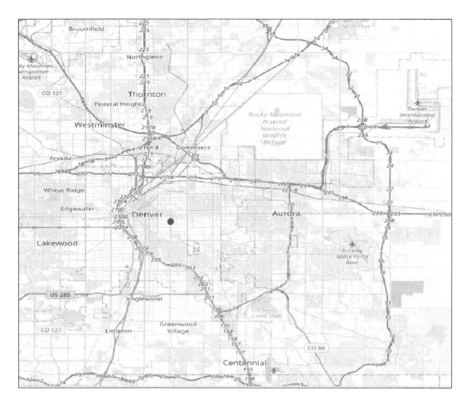

FIGURE 12.7 Screenshot of mean coordinate for all accidents in 2020. Basemap is the OpenStreetMap XYZ tiles which is © OpenStreetMap contributors and available under the Open Database License. Please see https://www.openstreetmap.org/copyright.

There should be a solitary point in the center of Denver, as shown in Figure 12.7. This represents the mean location for all accidents in the city, for which location data was explicitly provided. We return to this tool in a moment, but for now we also want to explore methods related to the distribution of the data. For a full appreciation of the statistical properties of the data, it is important to not only consider measures of the central tendency, but also the distribution. For the next two tools, we need to install plugins.

31. Navigate through the tab Plugins > Manage and Install Plugins.
32. Install both 'Standard Distance' and 'Standard Deviational Ellipse'.

First, we are going to calculate the standard distance. This should appear in the Processing Toolbox.

33. In the Processing Toolbox, scroll to the bottom to the newly added 'Standard Distance'.
34. Expand the tree twice to open the tool.

35. Set the Input Layer to Accidents2020_Complete.
36. Save the Output Layer to the GeoPackage as SDistance_All.
37. Keep the rest of the fields as default.
38. Ensure your dialogue box resembles Figure 12.8. Click Run.

We now have a standard distance around the mean center, as shown in Figure 12.9. This represents approximately one standard deviation of all the points, or approximately 68% of the spatial dispersion of the traffic accidents. The layers may need to be reordered in the Layers Panel.

The difference between the standard distance and the standard deviational ellipse is that the latter accounts for direction across the x and y axes. This means we can identify if there are any directional trends in the standard deviation of the data. There are a couple of options for accessing the standard deviational ellipse plugin tool that we recently installed. The Plugin Toolbar should now have a new Standard Deviational Ellipse button. If not, there should also be the option in the Vector tab.

39. Using either of the above methods, open the tool.
40. Again, select Accidents2020_Complete as the input (point) vector layer.
41. Ensure that Use Weights remains unchecked (for now).
42. Then select Yuill as the method.

A quick note, for most GIS operations, there are different methods that can be chosen. It is your job as the GIS user to select the most appropriate method for the task, but sometimes this can be daunting and quite complex. When we are tasked with such a choice, we can look to inbuilt support to provide guidance.

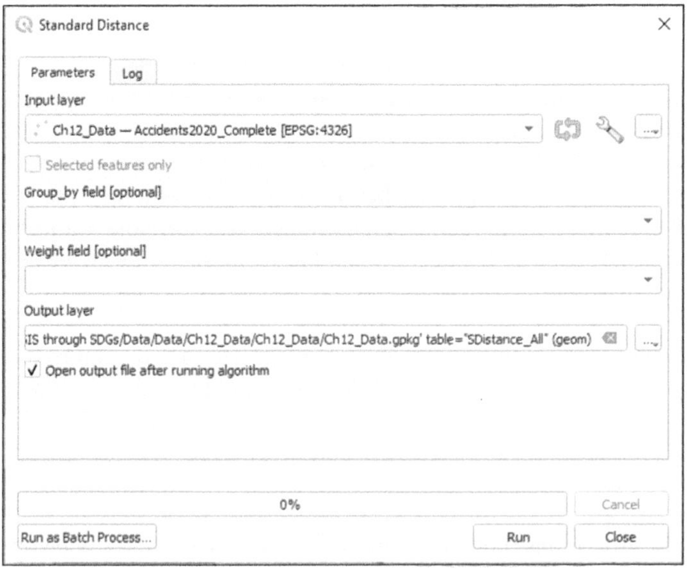

FIGURE 12.8 Screenshot of parameters for the standard distance tool.

Descriptive Statistics 213

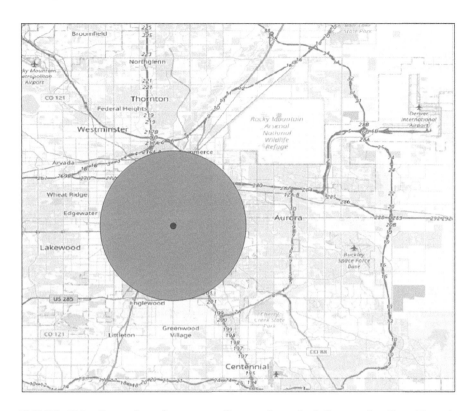

FIGURE 12.9 Screenshot of mean coordinate and standard distance for all accidents in 2020. Basemap is the OpenStreetMap XYZ tiles which is © OpenStreetMap contributors and available under the Open Database License. Please see https://www.openstreetmap.org/copyright.

43. Click on Help.

This should open a tab in the internet browser that provides information regarding the tool, which should outline the difference between Yuill and CrimeStat. Read up on the differences between the two methods. Once that is completed, we can progress with the tool.

44. Save the output vector layer as SDE_All (sde = standard deviational ellipse).
45. Ensure your dialogue box resembles Figure 12.10. Click OK.

Your map should now resemble Figure 12.11. There is quite a large geographic difference between the standard distance and standard deviational ellipse. The NE-SW directional shift in the ellipse indicates that the majority of the dispersion among accidents is occurring along the E-W axis as opposed to the N-S one. When we relate this to the aim of this case study to support SDG3.6, this is important as we may prioritize NE-SW routes as opposed to NW-SE routes in setting up mitigation efforts and awareness measures. However, we should treat this orientation with some

214 Understanding GIS through Sustainable Development Goals

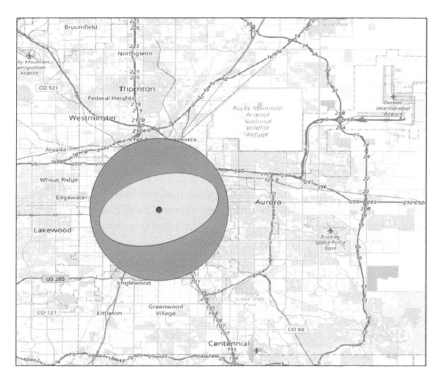

FIGURE 12.10 Screenshot of the parameters for the standard deviational ellipse tool.

FIGURE 12.11 Screenshot of the mean center, standard distance, and standard deviational ellipse for all accidents in 2020. Basemap is the OpenStreetMap XYZ tiles which is © OpenStreetMap contributors and available under the Open Database License. Please see https://www.openstreetmap.org/copyright.

FIGURE 12.12 Two examples of different road types, locations, light, and weather conditions in downtown Denver (right panel) and surrounding Colorado (left panel).

caution, given the overall spread of the data and the shape of the City of Denver administrative unit. If we turn on all the accident points, the administrative unit extends into the NW of the area toward the airport. Therefore, the spread along this E-W axis is a function of shape of the administrative unit more so than any underlying processes that appear to be occurring.

Therefore, to really examine the different spatial patterns of this dataset, we should aim to weight the accident points, as well as deconstruct the data based on the attributes. The attribute table of this dataset is very informative, including information on road conditions (dry, wet, etc.), light conditions (day light, night time, etc.), fatalities, serious accidents, and type of car. Therefore, we can use the attribute table to potentially glean more insight into the spatial patterns, and subsequently improve our understanding of the processes that might be shaping them. For example, Figure 12.12 demonstrates two different geographic locations in Denver and surrounding Colorado under different light conditions and weather conditions.

If certain features are more important than others, we can weight them. In our case study, the two variables that can best support SDG3.6 are serious injuries and fatalities. While there will always be extenuating factors that prevent serious injuries from becoming fatalities, and accidents from becoming serious injuries, we can consider these in our statistics. This could be particularly useful when we consider the target of SDG3.6 is to halve these serious and fatal accidents.

46. Open the mean coordinate(s) tool.
47. Select Accidents2020_Complete as the input layer.
48. Add Fatalities as the Weight field.
49. Save the layer in the GeoPackage, as MeanCenter_All_WF, with WF standing for Weighted Fatalities.
50. Ensure your dialogue box resembles Figure 12.13. Click Run.

Compare this to the unweighted mean center.

FIGURE 12.13 Screenshot of parameters for the mean coordinate(s) tool weighted by fatalities.

FIGURE 12.14 Screenshot of the mean center, standard distance, and standard deviational ellipse for all 2020 accidents unweighted and weighted by fatalities. Basemap is the OpenStreetMap XYZ tiles which is © OpenStreetMap contributors and available under the Open Database License. Please see https://www.openstreetmap.org/copyright.

51. Repeat the analysis for (1) standard distance and (2) standard deviation ellipse using Fatalities as the weighted variable. Do not forget to tick 'Use Weights' in the Standard Deviation Ellipse.

Your interface should now resemble Figure 12.14, although you may need to rearrange the layers in the Layers Panel to be able to visualize them all. There are some interesting trends emerging. When the statistics are weighted by fatalities, we observe

Descriptive Statistics 217

a strong eastern skew in the data, as well as a larger standard distance, which implies the fatal accidents are also more dispersed around the city, potentially in the surrounding areas as opposed to the city center.

52. Click on the Measure Line tool and measure the difference between the two mean centers. Turn on the snapping toolbar to use the magnet tool like we did in Chapter 4.

The difference is about 1130 m using the ellipsoidal distance, which suggests there could be a benefit in targeting our SDG3 mitigation efforts in the eastern part of the city. While the shape of the administrative unit will still influence the results, as the analysis is now weighted by fatalities, we can be confident there are factors that are causing higher fatalities in the east, which warrants further investigation. There may also be differences in the conditions that are present in different locations during these accidents. To investigate this, we deconstruct our dataset further using the attributes.

53. Open the Mean Coordinates tool again.
54. Select Accidents2020_Complete as the Input layer.
55. In the Unique ID Field, select LIGHT_CONDITION (this is the variable for light conditions).
56. Save the output in the GeoPackage as MeanCenter_All_UID (UID = Unique Identification Field).
57. Ensure your dialogue box resembles Figure 12.15. Click Run.

We now have several mean centers in the city, as shown in Figure 12.16. What QGIS has done here is that it has parsed the data based on the attribute value for light conditions at the time of the accident.

FIGURE 12.15 Screenshot of the parameters for the mean coordinates tool with a specific unique ID field.

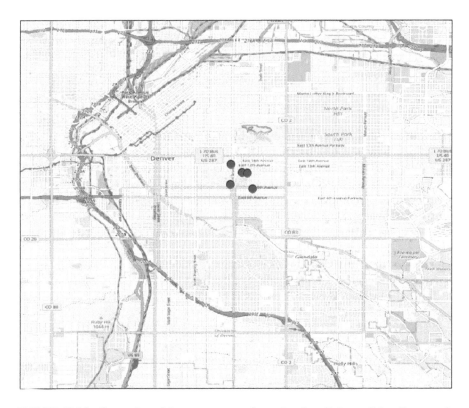

FIGURE 12.16 Screenshot of the mean centers for each unique light condition. Basemap is the OpenStreetMap XYZ tiles which is © OpenStreetMap contributors and available under the Open Database License. Please see https://www.openstreetmap.org/copyright.

58. Open the attribute table.

The attribute table should resemble Figure 12.17. Here we can see the mean x and mean y values for the different accidents that meet the specific criteria. The daylight condition is the furthest west out of all the points, meaning that the eastern skew may be in part contributed by accidents occurring in the suburbs during darker conditions. If we repeat this analysis, but weight the different conditions by fatalities, we can investigate how this impacts the spatial distribution.

59. Repeat these steps to specify the mean center, but this time add FATALITIES as the weighted variable alongside LIGHT_CONDITIONS as the Unique Field.

Observe how this changes the results, shown in Figure 12.18. The first thing to notice is that there are only four points instead of five. In this instance, there is one light condition (NULL) that does not have any fatalities. Interestingly, we appear to still have a more pronounced eastern skew for the two 'dark' conditions, but we have a much more southern skew for the dusk-dawn conditions.

Descriptive Statistics

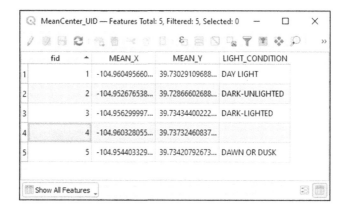

FIGURE 12.17 Screenshot of the attribute table for the mean centers for each unique light condition.

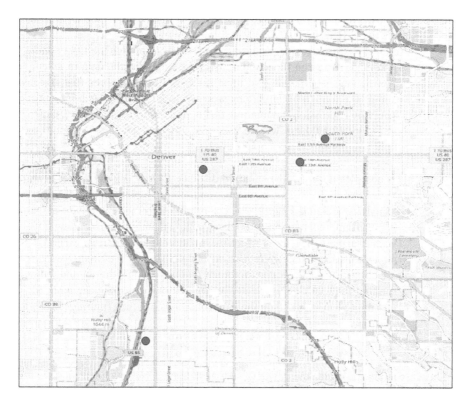

FIGURE 12.18 Screenshot of the mean centers for each unique light condition weighted by fatalities. Basemap is the OpenStreetMap XYZ tiles which is © OpenStreetMap contributors and available under the Open Database License. Please see https://www.openstreetmap.org/copyright.

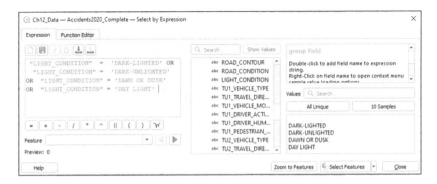

FIGURE 12.19 Screenshot of the select by expression to return all light conditions which have a valid attribute value.

We should always check whether patterns in dispersion are similar to those of central tendency.

60. Repeat this analysis for standard distance and standard deviational ellipse, but see the caveats below:

    Standard Distance – Because there are NULL values for the light condition field, if we run the standard distance tool with the data in its current format, it will fail. We must select only the features that match one of the four light conditions we have values for. As such, we must select by expression the features that match the following query shown in Figure 12.19. Here we are using the OR operator to select multiple light conditions.

Once this is completed, re-run the standard distance tool, which the Select Features Only option ticked.

    Standard Deviational Ellipse – There is no option in the standard deviational ellipse tool to parse through variables automatically, meaning if we want to explore the axis on which these patterns are occurring, we need to manually select them before running our analysis. This ultimately means selecting features by values for each unique light condition, and either running the tool on the selected values only or exporting the point file into multiple files based on the light conditions and then running the tool.

12.3 CASE STUDY CONCLUDING REMARKS

This case study illustrates the power of spatial statistics. We have identified patterns in the traffic accident data, specifically the center and spread of traffic accidents at a city-wide scale. We have weighted these accidents by fatalities, as well as deconstructing them based on the light conditions. Subsequently, we can use this analysis to reach an inference that fatal accidents at night occur in the east of the city, while during the day they are centered in the city. Furthermore, at dusk and dawn, key commuting times, there is an increase in fatal accidents in the south-west of the city. Such information allows identification of possible factors in the spatial patterns of accidents weighted by fatalities that are related to the traffic environment or drivers. This can then inform mitigation and adaptation measures, such as advert campaigns, increased visibility, and access to healthcare services, to be targeted toward such locations to reduce the number of fatal accidents (Sánchez-Mangas et al. 2010; Castillo-Manzano et al. 2014).

The three new tools used in this chapter, mean coordinates, standard distance, and standard deviational ellipse, have provided new insights into this dataset. Moreover, by using the attribute table, we have successfully weighted the accidents such that patterns of more serious incidents can be discerned, as well as deconstructing (or subsetting) the dataset to research how changes in light conditions influence the spatial patterns. Finally, we have reinforced our knowledge of the attribute table, performing more advanced attribute queries using Boolean logic.

12.3.1 Test Yourself

If you want to test yourself on the learning outcomes of this chapter, complete the following:

a. Repeat the analysis to investigate whether the same patterns are observed when the analysis is weighted using serious accidents instead of fatalities.
b. Can you use the skills learnt in Section 3 to create a map in the Print Composer that summarizes the key information you have generated in this chapter?

REFERENCES

Castillo-Manzano, J.I., Castro-Nuño, M. and Fageda, X., 2014. Can health public expenditure reduce the tragic consequences of road traffic accidents? The EU-27 experience. *The European Journal of Health Economics*, 15(6), pp. 645–652.

Lek-Uthai, U., Sangsayan, J., Kachenchart, B., Kulpradit, K., Sujirarat, D. and Honda, K., 2010. Novel ellipsoid spatial analysis for determining malaria risk at the village level. *Acta Tropica*, 116(1), pp. 51–60.

McGrath, S.A., Perumean-Chaney, S.E. and Sloan III, J.J., 2014. Property crime on college campuses: A case study using GIS and related tools. *Security Journal*, 27(3), pp. 263–283.

Rogerson, P.A., 2020. *Statistical Methods for Geography: A Student's Guide*. Sage.

Sánchez-Mangas, R., García-Ferrrer, A., De Juan, A. and Arroyo, A.M., 2010. The probability of death in road traffic accidents. How important is a quick medical response? *Accident Analysis & Prevention*, 42(4), pp. 1048–1056.

13 Density

13.1 INTRODUCTION AND LEARNING OUTCOMES

Density analysis identifies where features are concentrated. We can represent a multitude of features in GIS as points (i.e., shops, trees, and crime) and lines (i.e., roads, rivers, and fences), and we can use density analysis to identify where the highest concentration of these features are. By using GIS to identify locations that consist of high or low densities of phenomena, we can support informed decision making regarding specific application areas. In this chapter, we generate density maps and models of crime.

Using GIS to support criminology has a long history and is certainly not unique to the SDGs. However, the SDGs include several targets that are strongly related to a reduction and prevention of crime. SDG16 aims to promote peaceful inclusive societies, with target SDG16.1 to reduce all forms of violence and related death rates; SDG16.2 to end abuse, exploitation, trafficking, and violence against children; and SDG16.4 to reduce and combat all forms of organized crime. Similarly, SDG5 on Gender and Equality, specifically target SDG5.2, aims to eliminate all forms of violence against women and girls in the public and private spheres. Building on one of the themes of this book, if we know where crimes are occurring in high densities, we can incorporate preventative measures to ensure a reduction in such events. Moreover, the spatial pattern may identify factors that compound the issues, potentially increasing the opportunity for educational or social reform in high-density areas or learn from applied policies in low-density areas (Zhou et al. 2014).

In this chapter, we use data from Austin, Texas, USA. I chose Austin as a case study as the Austin Police Department (2016) published location data of crimes for 2015 in the Public Domain, and it's an area I'm familiar with having lived (very safely I might add!) from 2011 to 2016. In 2015, there were 38,573 crimes reported, with categories pre-processed into violent crime, theft, auto-theft, and burglary. We initially explore the dataset using density analysis across all crime types, before focusing on specific crimes.

By the end of this chapter, you will have completed three learning outcomes, and you should be able to:

- Write a CASE statement to generate a new attribute
- Visualize the point data using a heatmap
- Implement density analysis for specific crimes by subsetting the attribute table

13.2 CASE STUDY: MODELING CRIME HOTSPOTS IN AUSTIN, TEXAS, USA, TO SUPPORT THE SDGs

1. Open a New Empty Project in QGIS.
2. Save the project as Ch13.

3. Navigate through the tab Layer > Add Layer > Add Delimited Text File.
4. This layer has geometry provided. The CRS system is a NAD 83 State Plane Texas Central FIPS 4203, with the measurement unit feet. FYI, this is EPSG code 102739.
5. Click OK to approve any transformations.

One quick note, this projection reports the linear units in feet, which is different to the predominant metric system we have used throughout the book so far.

6. Turn on the OSM basemap using the Browser Panel and navigate to XYZ tile to ensure the data is loaded in the correct place.

Your map should resemble Figure 13.1. The data is relatively spread across the city, which is expected with all crime categories represented in a single point layer. The first thing we do when considering the density of a layer is to use cartography to represent density through a heatmap. Currently the data is represented using a single symbol, a point. We can visualize the data as a heatmap in the symbology tab, again building on the skills learnt in the previous cartography section.

7. Right click on Ch13_Data layer to open the properties, and navigate to the symbology tab.
8. Click on Single Symbol and change this to Heatmap.
9. Click on the dropdown menu to change the color ramp from Black to Reds.

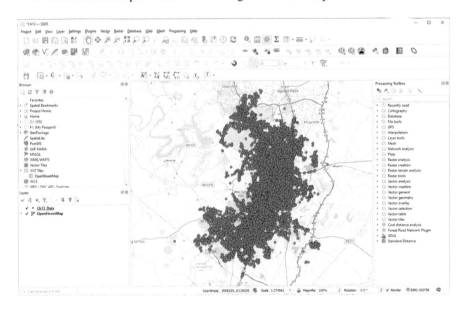

FIGURE 13.1 Screenshot of the point representation of the Austin Police Department (2016) crime data in Austin, Texas, USA. Basemap is the OpenStreetMap XYZ tiles which is © OpenStreetMap contributors and available under the Open Database License. Please see https://www.openstreetmap.org/copyright.

Density   225

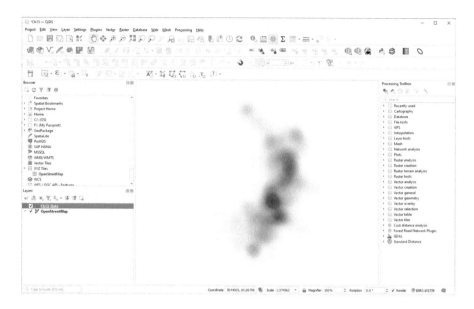

FIGURE 13.2 Screenshot of the default heatmap symbology for the crime data.

   10. Keep the rest of the options as default for the moment but note that we will return to change the radius.
   11. Click Apply.

We have generated a heatmap that displays darker colors for higher density areas, as shown in Figure 13.2. We can see the darkest area in the center of the map, with other hotspots present throughout the city. However, this is a rendering. This means if we zoom in and out, the heatmap updates and changes.

   12. Zoom in and out on the data to see how the hotspots change.

When we zoom in, we get more localized hotspots than when we view the data at a city-wide scale. However, it is difficult to link the heatmap with the basemap as the heatmap layer is fully opaque (or non-transparent). Luckily, we can change that.

   13. Reopen the properties for the layer and navigate to symbology.
   14. Click on layer rendering, just above style in the bottom left of the dialogue. This should open more options.
   15. Change the opacity to 50%. This is equivalent of setting the layer to be 50% transparent.
   16. Ensure your dialogue box resembles Figure 13.3. Click Apply then OK.

Now we can see both the hotspot visualization of the crime data as well as the basemap. This allows us to interrogate the data with geographic context. Pan and zoom around the city and look for patterns. The highest density of points is in the

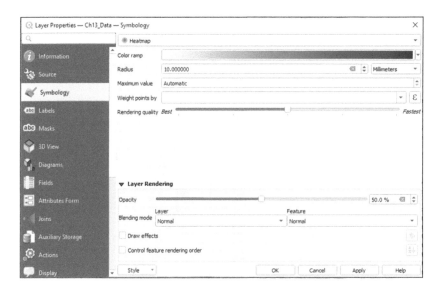

FIGURE 13.3 Screenshot of parameters to update opacity in heatmap symbology.

downtown region, while there are several hotspots aligning with shopping malls and car parks. These locations represent a single X and Y location that has several crime features, partly due to the geographic nature of the infrastructure. For example, all crimes reported at a shopping mall have the same X and Y coordinates.

Visualizing point data like this is useful; however, with the changing render it does mean that as we move around the map, the hotspots change. This is because we used the default settings in the visualization process. Again, as will become a common theme of the book, when working with GIS we need to consider that the default settings may not always be the best option for our work.

17. Reopen the properties tab and navigate to symbology.

Notice that the radius was set to 10 mm. This parameter determines the neighborhood around each point that is considered in calculating density. The default setting of 10 mm means that as we zoom in and out, the number of points that are used in the density algorithm changes to all points within 10 mm on the map (i.e., not real-world units). Subsequently, the results change too. As we know from Chapter 11, the size of the neighborhood has a large influence on the output of the results, and again this is no exception. While such an adaptive method is great for exploring the data, sometimes it is best to have a consistent value that does not change as we pan and zoom.

18. Click on the dropdown of Millimeters.

Here we can see various measures of distance and adjacency (i.e., neighborhoods) that can be used in the density algorithm.

19. Select Map Units.
20. As the map measures distance in feet, use a value of 6561. This is equivalent to 2 km.
21. Click Apply and OK.

The number of hotspots has decreased, and as we zoom around the map, these locations do not alter. When working with neighborhoods if we increase the range of influence, the smoother our data and relationships become. This is due to the wider geographic sphere of influence and the fact that closer locations tend to be more similar than distant ones, the basis for Tobler's First Law of Geography (Tobler 1970). Therefore, if we change the neighborhood distance, we get different results.

22. Reapply the heatmap symbology using a radius of 3280 feet (equivalent to 1 km).

With a smaller neighborhood defined, we now have more hotspots, as can be seen in the comparison of the two outputs shown in Figure 13.4. We could further advance this work by weighting crimes by their severity, which would involve creating a new attribute in the layer and writing a CASE statement to assign different values based on different conditions.

23. Open the attribute table, and select Field Calculator.
24. Name the new field 'Weight' and set it to Whole number (integer).
25. Write the following expression in the Field Calculator to create a new weight attribute that ranks violent crime as 10 and other crimes as 1.

```
CASE
   WHEN "Offense Description" = 'Violent Crime' THEN 10
   ELSE 1
   END
```

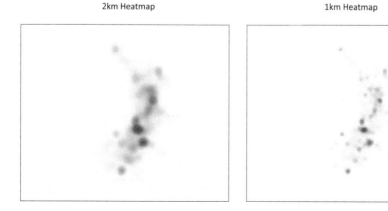

FIGURE 13.4 Comparison of heatmaps using a 2 km and 1 km radius.

26. Click OK.
27. Open the attribute table to confirm this has worked. If not, check the CASE statement.

The case statement is the equivalent of IF ELSE logic. IF the value for Offense Description is 'Violent Crime', then the new value for Weight will be 10, ELSE it will be 1.

28. Return to the symbology tab in properties and repeat the previous steps to visualize the data using a heatmap; however, this time weight is using our new Weight attribute.

Your dialogue box should resemble Figure 13.5.

The hotspots in downtown Austin persist, as shown in Figure 13.6, but those along the main road (I35) running N-S in the city disappear. By weighting violent crime ten times that of other crimes, it could support organizations working toward SDG16.1, as there is an increased understanding of where violent crimes are occurring. We discussed and implemented weighting in detail in the previous chapter, so the main aim of this exercise was to demonstrate how to write a CASE statement to generate a weight field. This could be applied to any of the three other offense types should the focus of the analysis warrant exploration of a different type of crime.

The other limitation of simply visualizing this data is that the hotspot representation is temporary. Or in other words it is not its own spatial layer that we can incorporate into subsequent spatial analysis. This is important as we progress through the

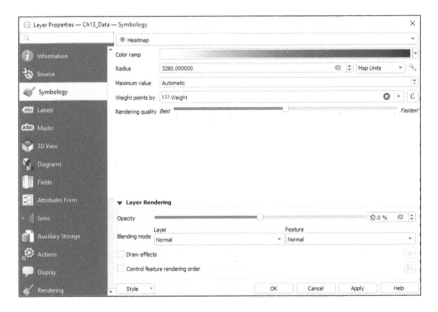

FIGURE 13.5 Screenshot of the parameters to create a heatmap symbology with an increased weight on violent crimes.

Density

FIGURE 13.6 Hotspot of crimes in Austin, Texas, USA, in 2015 weighted by violent crimes. The heatmap has a 50% transparency, and is overlain on a basemap. Basemap is the OpenStreetMap XYZ tiles which is © OpenStreetMap contributors and available under the Open Database License. Please see https://www.openstreetmap.org/copyright.

book where we use the analytical methods outlined in this section to generate spatial layers that can subsequently be used in more complex analytical models. The remainder of this chapter demonstrates how to generate a permanent raster layer of density.

29. Change the symbology of the crime data (Ch13_Data) back to Single Symbol points.
30. Click Select Features by Values.
31. Set Offense Description to equal 'Auto Theft'.
32. Ensure your dialogue box resembles Figure 13.7. Click Select Features.
33. Right click on the crime layer and select Export > Save Selected Features As....
34. Save the layer as Auto_Crime within a new GeoPackage called Ch13_Data.
35. Click OK.

This new layer represents all auto-theft crime locations. The organized theft of motor vehicles has immediate financial and safety concerns to the individual owner, but also implications for insurance companies, car manufacturers, and in most cases

FIGURE 13.7 Screenshot of the query to select all auto-thefts from the crime dataset.

these thefts are linked to organized crime (INTERPOL 2022). Therefore, identifying areas of high-density auto-theft could support SDG16.4, through increased awareness of these crimes using warning signage to promote preventative measures (i.e., immobilizers, GPS trackers, and Faraday cages). It could also support strategic placement of CCTV and patrols, which could lead to a disruption and prevention of organized crime.

For density analysis in QGIS, we use kernel density to generate a heatmap. The kernel density method calculates hotspots and clustering based on the number of points in a neighborhood. The more points within a specified area, the higher the density. The analysis works in a similar fashion to the visualization, but this tool provides more options when specifying the neighborhood. The heatmap visualization uses a Quartic function to generate the kernel shape when defining the influence of other points within the neighborhood, but there are various kernel shapes that can be fit.

The tool Heatmap (Kernel Density Estimation) calculates this using probability density functions. A theoretical curve estimates density as the probability that a specific value is found when distributed over all values of the variable. This can be considered as the relative number of times we expect a random variable to assume each and every possible value. Because a continuous variable has infinite possibilities of values, we use the density function to calculate the probability of a value through calculus. A normal distribution has a bell-shaped curve that centers on the mean value and spreads out so that 95% of the data points are within 1.96 standard deviations of the mean. Again, this builds on the information from the previous chapter where we calculated standard distances that were capturing the standard deviation of the spatial points. While it is not the aim of this book to revisit the statistical methods used, it is important to have a basic understanding of how we calculate spatial density. Kernel Density Estimation uses this density function and transforms it so that it can be represented in space, meaning each value returned from the analysis represents the probability density function at that specific location.

36. In the Processing Toolbox, search for Heatmap within the Interpolation tools.
37. Specify the point layer as AutoCrime.

Density

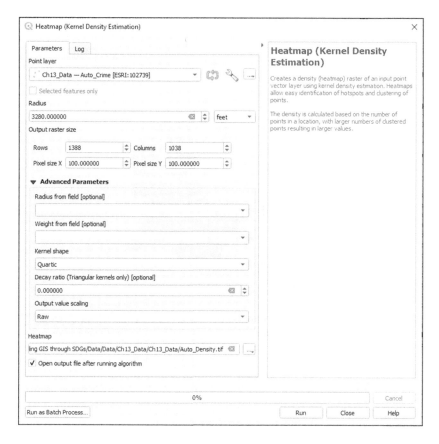

FIGURE 13.8 Screenshot of the parameters needed for the kernel density estimation tool.

38. Set the radius to 3280 feet.
39. Change the pixel size to 100.
40. Keep the kernel shape as Quartic.
41. Save the output layer to your working directory as Auto_Density.
42. Ensure your dialogue box resembles Figure 13.8. Click Run.

This creates a raster layer that represents the density of auto crimes, as shown in Figure 13.9. From the Layers Panel, we can see that the density values range from 0 to 43.38. The density is calculated within the radius specified in the above tool (3280 feet), with the influence of points captured using a Quartic kernel shape. The second thing we should notice is that this layer does not cover the entirety of the city, with locations beyond 3280 feet of any auto crime given a NoData value.

NoData represents locations for which we have no information or data. There is a marked distinction between 0 and NoData. Zero is a unique value while NoData is an acknowledgment that we do not have information for that location. Therefore, we should never discount the meaning of not having data in a specified area. While our data is comprehensive across the city, the city limits are not uniform, meaning any

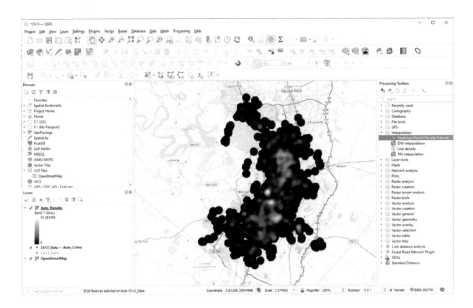

FIGURE 13.9 Screenshot of the Auto_Density layer. Basemap is the OpenStreetMap XYZ tiles which is © OpenStreetMap contributors and available under the Open Database License. Please see https://www.openstreetmap.org/copyright.

filling of NoData values will result in inaccurate zero values where we could have crimes, just reported in a different jurisdiction. Therefore, we keep the data in its current format to prevent any misinterpretation.

Next, we want to symbolize this more effectively to portray the key information. Currently, this is displayed as a singleband gray render, with lighter colors representing a higher density. We can change the color palette to match that used at the start of the chapter.

43. Navigate to Properties > Symbology.
44. Change singleband gray to singleband pseudocolor.
45. Change the color ramp to Reds.
46. Click Classify to populate the categories of values. In this instance, we are using the continuous mode of categorizing.
47. Ensure your dialogue box resembles Figure 13.10. Click Apply.

We also want to change the transparency of the raster layer so that we can see the basemap for geographic context. As the layer is a raster, this is different to the function used for vector data. In the properties dialogue box, there should be a tab called Transparency, immediately below the symbology tab.

48. Click on the Transparency tab in the Properties dialogue box.
49. Change the Global Opacity to 70%.
50. Click Apply.

Density   233

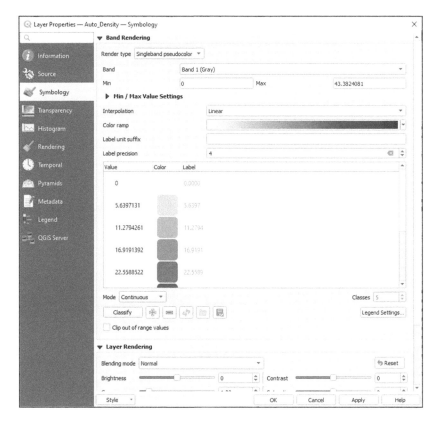

FIGURE 13.10   Screenshot of parameters to symbolize the raster layer Auto_Density.

If the basemap is not visible, turn the remaining layers off. Explore how the hotspots correspond to the overall crime locations.

13.3   CASE STUDY CONCLUDING REMARKS

We can see there remains an area of high-density downtown. This is an area where there is a lot of night life, with a high density of bars and tourists. There are also two new areas of high density that appear close to the city center. By using the zoom and pan tools we can identify that these are in East Riverside-Oltorf and near West 24th Street. These two areas are locations where there is a high college student population and purpose-built student accommodation. While nuanced, the in-migration of students into a neighborhood is often associated with an increase in auto crime, in part due to the possession of expensive vehicles and items, coupled with a lack of sufficient security or protection (Visser & Kisting 2019). We can also observe other areas of high density close to the airport and in the north of the city along the 240 and 183 Toll road, with such locations representing a different type of land use, but largely areas where there is a high turnover of residents and vehicles in areas where social tourism and transient populations aggregate, which have been associated with higher

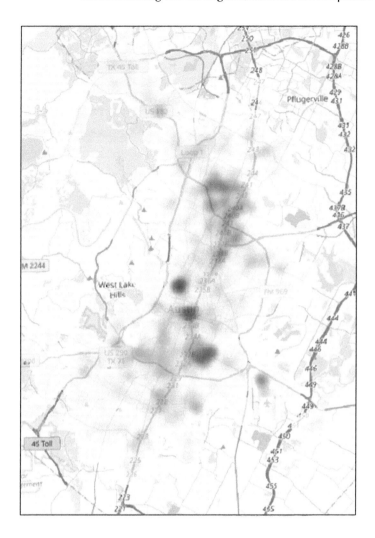

FIGURE 13.11 Screenshot of the heatmap for auto-theft in Austin, Texas, USA. Basemap is the OpenStreetMap XYZ tiles which is © OpenStreetMap contributors and available under the Open Database License. Please see https://www.openstreetmap.org/copyright.

crime rates, including auto-theft (Xu et al. 2019; Roth 2020). By deconstructing the data based on attributes and analyzing where such crimes occur in their highest density, targeted prevention and education measures can be established.

Using spatial analysis and cartography, we have identified geographic patterns in the spatial distribution of crimes in Austin. We used the heatmap renderer, observing how clusters are identified, as well as the need to consider parameters beyond the default settings. We then used field calculator to write an advanced CASE statement to generate a new attribute, which we then used to weight the visualization. Finally, we used Kernel Density Estimation to generate a permanent density layer, focusing specifically on auto crimes. In the next chapter, the final one of this section, we explore more interpolation methods to convert point data into a continuous representation of space.

13.3.1 Test Yourself

If you want to test yourself on the learning outcomes of this chapter, complete the following:

a. Can you change the kernel shape in the heatmap tool and investigate how this impacts the results?
b. Can you write a CASE statement to include different crime offenses in the weighting and generate a new density output?

REFERENCES

Austin Police Department (2016). *Annual Crime Dataset 2015*. Available from: https://data.austintexas.gov/Public-Safety/Annual-Crime-Dataset-2015/spbg-9v94. Accessed March 1, 2022. https://doi.org/10.26000/043.000003.

Interpol (2022). *Vehicle Crime*. Available from: https://www.interpol.int/en/Crimes/Vehicle-crime#:~:text=The%20organized%20theft%20of%20motor%20vehicles%2C%20while%20of,%E2%80%93%20is%20linked%20to%20other%20organized%20crime%20operations.

Roth, J.J., 2021. Home sharing and crime across neighborhoods: An analysis of Austin, Texas. *Criminal Justice Review*, 46(1), pp. 40–52.

Tobler, W.R., 1970. A computer movie simulating urban growth in the Detroit region. *Economic Geography*, 46(sup1), pp. 234–240.

Visser, G. and Kisting, D., 2019. Studentification in Stellenbosch, South Africa. *Urbani izziv*, 30, pp. 158–177.

Xu, Y.H., Pennington-Gray, L. and Kim, J., 2019. The sharing economy: A geographically weighted regression approach to examine crime and the shared lodging sector. *Journal of Travel Research*, 58(7), pp. 1193–1208.

Zhou, G., Lin, J. and Ma, X., 2014. A web-based GIS for crime mapping and decision support. In *Forensic GIS* (pp. 221–243), Elmes, G., Roedl, G., & Conley J (Eds). Springer, Dordrecht.

14 Interpolation

14.1 INTRODUCTION AND LEARNING OUTCOMES

Throughout this book, we have used a variety of datasets that represent real-world geographic features using both discrete and continuous representations. So far, most of the datasets used have consisted of the vector spatial data model. We know from our experience thus far, however, that certain phenomena may not be best represented using a discrete conceptualization of space. Despite that, not all data is captured in the format that best suits its analysis or visualization. In fact, a lot of data on continuous features is collected at discrete locations. For example, weather stations capture the meteorological conditions across a series of explicit locations and geological cores measure the rock type at specific locations. This data captured at these discrete locations represents as close to the true conditions as possible, but they do not inform on areas in-between. Therefore, we need a method to fill in the gaps.

We use interpolation to estimate missing values within the range of sample data. Consider a simplistic elevation profile, as demonstrated in Figure 14.1, where we have elevation values every 100 m (or any distance for that matter). However, upon recording this information, our instrument fails to record the specific value at point C. We can use simple linear interpolation to assume that the value falls between those of B and C. If we consider B has an elevation of 4 m and C has an elevation of 12 m, C is simply the sum of 4 and 12, divided by 2, which makes the elevation of point C 8 m. This is the most simplistic representation of interpolation, and there are various non-linear equivalents we could use. However, the concept is a useful analogy to introduce interpolation, and while this example occurs in one-dimensional space, when we interpolate in GIS we do so in two-dimensional space.

Spatial interpolation therefore works on the same principles. It is the process of predicting the values of any attribute (i.e., weather, elevation, and geology) at unsampled sites from sampled locations. It is fundamentally based on mathematical functions that incorporate the values of the sample points and their distance from intended interpolated points. This is not to be confused with extrapolation, which when dealing with spatial data means estimating a value beyond the known value range or spatial extent of our sample data.

Spatial interpolation is useful when our aim is to convert discrete points to a continuous representation of space. There are various interpolators that can be used, that each demand different sample sizes and mathematical functions. It should be noted here that there is no dominant interpolation method, and that we should choose our method based on the specific purpose of our study and different aspects of our data and methods. However, the one thing in common across all interpolation methods is the underpinning law that allows such quantification.

Tobler's First Law of Geography (Tobler 1970) is perhaps the most important concept in geography and GIS, as it provides a quantitative framework to estimate

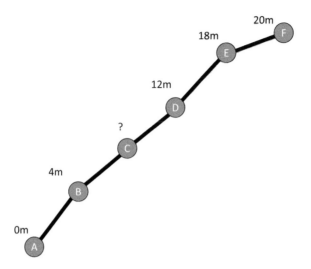

FIGURE 14.1 Example of simple linear interpolation in one-dimensional space.

values in space, as well as test hypotheses and create new data. The law simply states, when two values are near each other, they are more likely to be similar to each other than distant values. Or in other words, as the distance between two observations increases, the similarity between the attributes of those observations decreases. Our non-spatial example provides a simple illustration of this, as it is unlikely that such an elevation value between two locations is going to be much higher or lower. As with most spatial analysis, we are undertaking this intuitively when we first observe data.

This chapter explores interpolation through SDG7 Affordable and Clean Energy. Given the role natural processes, specifically those related to climate and weather (i.e., wind and solar insolation), will play in ensuring a larger reliance on renewable energy sources, we focus on how meteorological stations and interpolation can be used to support SDG7.1 and 7.2. These targets aim to ensure universal access to affordable, reliable, and modern energy services, as well as increasing the share of renewable energy within the global energy mix. By the end of this chapter, you will have completed two learning outcomes, and you should be able to:

- Perform spatial interpolation using two different methods
- Critically explore the impact of the neighborhood on interpolation methods

14.2 CASE STUDY: SDG7 INTERPOLATING SOLAR RADIATION IN CHINA

Here we use a dataset developed by Tang (2019), which contains daily average solar radiation of 716 weather stations in China spanning from 1961 to 2010 see also Tang et al. (2010; 2013). In raw format, the data themselves are presented as 717 text files, with one file representing the spatial location of each station and 716 files of daily solar radiation licensed under an Attribution 4.0 International (CC-BY-4.0).

Interpolation

Ch14_Data consists of one *.csv file that includes the spatial location of the 642 weather stations that have recorded solar radiation for 2010. Within this dataset there is the annual mean and standard deviation of solar radiation for each weather station. With such a detailed dataset, the data could have been captured on a yearly, monthly, weekly, or even daily timescale to inform optimal energy efficiency questions. Given the overall learning outcomes related to interpolating data from discrete points to a continuous surface, using a yearly statistical summary should support GIS understanding as well as a thematic application.

1. Open a New Empty Project in QGIS.
2. Add both layers from Ch14_Data GeoPackage.
3. Save the project as Ch14.

Your project should resemble Figure 14.2. There is one point layer representing the 643 weather stations with solar radiation data for 2010 from Tang (2019) and an outline of China that has been exported as its own layer from the world-administrative-boundaries file obtained from the World Food Programme (2019) licensed under an Open Government License v3.0. The point layer does not cover the entirety of the polygon layer, with a large area in the west not having any weather stations. Our task is to convert the information within the point dataset to a continuous representation using interpolation methods. In QGIS, there are two in-built methodologies that we focus on in this chapter.

Triangulated Irregular Network (TIN) – The TIN Interpolation function in QGIS creates a surface formed by triangles of nearest neighbor points. Each sample point has a circumcircle generated around it, with their intersections connected to a network of non-overlapping triangles. These triangles are generated so that they are as compact as possible.

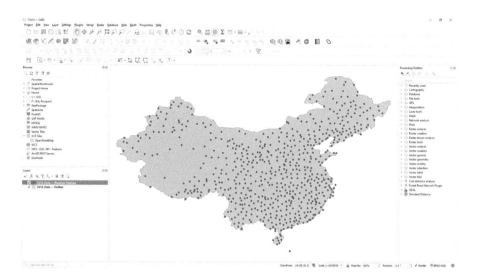

FIGURE 14.2 Screenshot of the weather station data and outline of China.

Inverse Distance Weighting (IDW) – The IDW interpolation function in QGIS generates a continuous representation by weighting each sample point as the inverse proportion of the distance weight. It accounts for nearness by selecting sample points either within a kernel radius, or as a minimum number of 'near' points. Again, think back to Chapter 11 when we discussed what 'near' actually means. Subsequently, changing the radius or number of near points will affect the resultant surface, meaning we should examine how changing these values impacts our work. In general, closer values become more influential as the sample size increases, while the higher number of points specified as 'near' results in a smoother surface. This is the same process as we explored in the previous chapter when we decreased the neighborhood size in the heatmap visualization.

4. In the Processing Toolbox, navigate to Interpolation.

Within this set of tools, we should see IDW interpolation and TIN interpolation.

5. Double click on TIN interpolation to open the tool.
6. Specify the Weather Station points as the Vector Layer.
7. Change the interpolation attribute to Mean.
8. Click the Green + sign, which adds this value to the interpolation function.

We could add more attributes using this function, which creates multi-band rasters with multiple interpolations; however, for now we keep the interpolation straightforward and keep the two values (mean and standard deviation) separate.

The next 'option' we can change is the interpolation method. There are two options, linear or cubic. This is the equivalent of a linear interpolation explained in Figure 14.1, or a non-linear representation of this that uses a cubic method.

9. Choose linear.
10. Specify the extent as the same as the outline layer.
11. Save the output in the Ch14 extracted folder as 'TIN_Mean_Linear'.
12. Ensure your dialogue box resembles Figure 14.3. Click Run.

A new raster layer appears in the project, as shown in Figure 14.4. There are a few things to notice. Firstly, think back to the definition of interpolation, which means 'filling in the gaps'. Values have been created for each x and y grid between the existing points. However, this has been completed within a minimum bounding polygon of all points, meaning that on the outskirts of the country there are areas that are missing. If we wanted these locations to be completed, we would be undertaking a hybrid form of interpolation and extrapolation. In other GIS software, this can be overcome, but in QGIS, there is no way to extrapolate the TIN to cover the entirety of the extent. Regardless, even when this is possible, values beyond these locations become increasingly unrealistic.

Secondly, we should see the triangular nature of the raster, and if we zoom in, each grid has a unique value indicative of a continuous representation. However,

Interpolation

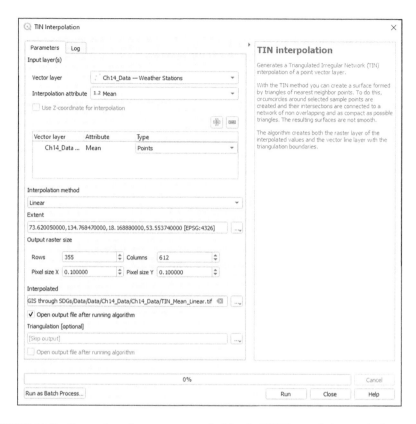

FIGURE 14.3 Screenshot of parameters needed for the TIN Interpolation method.

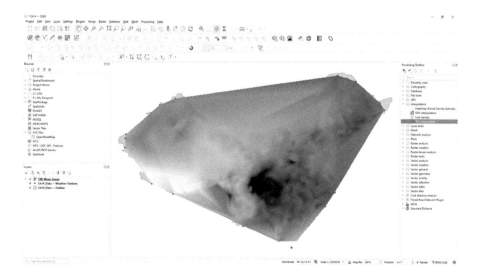

FIGURE 14.4 Screenshot of completed TIN interpolation for mean solar radiation.

the representation of solar insolation appears quite blocky. At such a spatial scale, this triangular representation might not be wholly appropriate for representing solar insolation. Next, we explore the IDW method.

13. Double click on IDW interpolation in the Processing Toolbox.
14. Again, select the weather stations as the vector layer, select the mean as the attribute value, and click the green +.

Next, we must select a 'Distance coefficient P'. The distance coefficient is how the impact of neighboring points is defined. A larger coefficient will reduce the impact of points further away.

15. For the moment keep the default value of 2.
16. Select the Extent as the outline of China.
17. Save the output layer to your working directory as IDW2_Mean.
18. Ensure your dialogue box resembles Figure 14.5. Click Run.

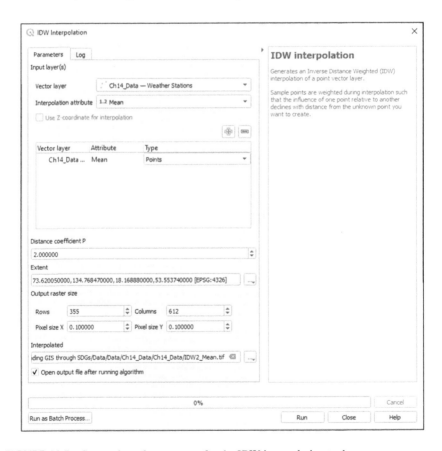

FIGURE 14.5 Screenshot of parameters for the IDW interpolation tool.

Interpolation

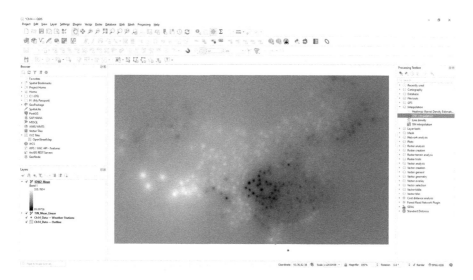

FIGURE 14.6 Screenshot of IDW with distance coefficient P 2.

A new layer is created that should resemble Figure 14.6. Firstly, we should notice that the extent is now in-line with that of the overall extent of the study area. To match the boundary exactly, we need to clip the raster by the mask.

19. Navigate through the tab Raster > Extraction > Clip raster by Mask Layer.
20. Select IDW2_Mean as input layer.
21. Select Outline as the mask layer.
22. Specify WGS84 as the CRS.
23. Save the output as IDW2_Mean_Clip.
24. Click Run.

A new layer is generated that is now at the extent of China, as shown in Figure 14.7. We must turn off the original IDW2_Mean layer in the Layers Panel to observe this. The next point to notice is that this map looks quite spotty, with the values of the raster becoming smoother as the distance away from the points increases. This is a direct result of the coefficient, as smaller coefficient values mean that the impact of closer values is weighted higher. Let us repeat this process with a larger P.

25. Repeat steps 13–24, but this time change the Distance Coefficient P to 50, and save the layers as IDW50_Mean and IDW50_Mean_Clip.

We should observe an immediate difference between the two methods, with the results of the Distance Coefficient 50 shown in Figure 14.8. This output looks much blockier, in fact it looks more like a set of polygons, but in raster format. If we zoom to one of these blocks and use the Identify Features tool, we see that the values are largely the same. Think carefully about the different inputs, outputs, and weighting options.

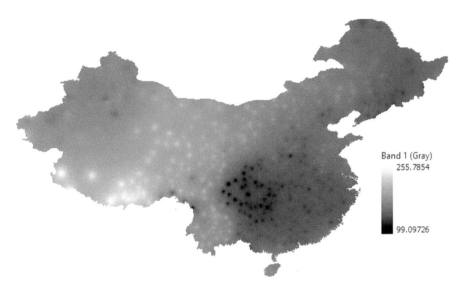

FIGURE 14.7 Screenshot of IDW with distance coefficient P 2 clipped to China.

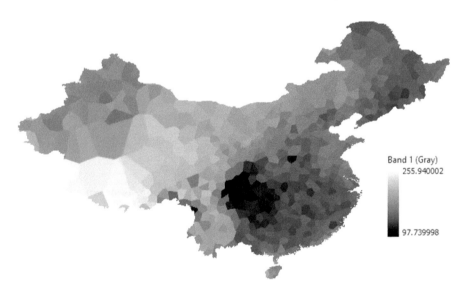

FIGURE 14.8 Screenshot of IDW with distance coefficient P 50 clipped to China.

26. Try using the IDW interpolation tool with different values of P. For example, use 3, 25, etc. Compare the outputs. This allows us to observe whether different artifacts have been introduced by the chosen methodology or parameters.

One helpful hint is to try and develop a coding system that allows us to name our new solar insolation rasters (e.g., IDW12 for a coefficient distance P of 12). In my experience of teaching GIS over several years, when naming conventions in instructions

Interpolation

are provided, they are not always followed. However, given the ability of QGIS to store temporary files with the name of the GIS function, one could quite easily end up with a Layers Panel of several 'interpolated' layers, which makes interpretation and understanding of which method might be the most suitable difficult.

Visual analysis of these methods is useful but does not quantify where the largest differences in interpolated surfaces are. In GIS, we can overlay rasters using arithmetic operators using the raster calculator to compare the two IDW interpolations we created earlier.

27. Navigate through the tab Raster > Raster Calculator.
28. Double click on IDW2_Mean_Clip, then click on the '−' sign, and finally double click on IDW50_Mean_Clip.

This expression subtracts the solar insolation values generated by IDW50 from IDW2. Where the values align perfectly, the output will be 0; where the values are positive, IDW2 has created a larger value of solar insolation than IDW50, and when negative the opposite has occurred.

29. Save the output in your working directory, as a *.tif file called difference.
30. Ensure your dialogue box resembles Figure 14.9. Click OK.

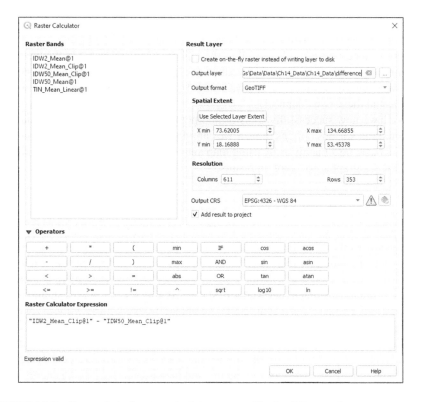

FIGURE 14.9 Screenshot of raster calculator to quantify the difference between outputs.

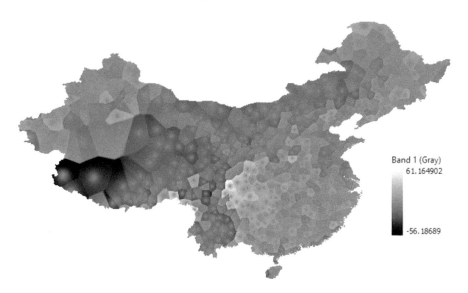

FIGURE 14.10 Difference between the IDW2 and IDW50 methods.

The resultant raster output should resemble Figure 14.10. We can see that in the east of the country, our values are largely positive (white), meaning IDW2 has estimated a larger insolation, while in the west of the country the values are negative (black), meaning IDW50 has estimated the larger insolation. Given the blocky nature of the larger distance coefficient P, as well as the higher values estimated in the west of the country where we have less observations, we can now make an informed decision that we intend to advance our analysis with IDW2 (on the assumption that we have only tested these two distances).

We have now successfully transformed point data into a continuous surface and explored the various options of how we can implement this. Moreover, we have performed basic raster analysis by using map algebra to compare the outputs and examined the geographic variation in results.

14.3 CASE STUDY CONCLUDING REMARKS

We can see that the highest annual mean solar insolation across the country is predominantly in the west, which would correspond with the fact that this area is a desert. Therefore, based on solar insolation, this area would perhaps be the best location for a large-scale solar photovoltaic farm (Safriel 2009). However, there is increasing research suggesting that to best support SDG7, existing rooftop spaces should be used to install solar panels (Joshi et al. 2021). Therefore, we could overlay these solar insolation interpolations onto a population density or building density map and identify the locations with the highest mean solar insolation and building density. Such steps would represent perhaps one of the simpler methods of governments identifying where best to provide solar energy infrastructure and increase its share in the energy mix.

The main aim of this chapter was to undertake interpolation, and subsequently while we have begun to address SDG7, we have done so in a very simplistic manner.

Interpolation

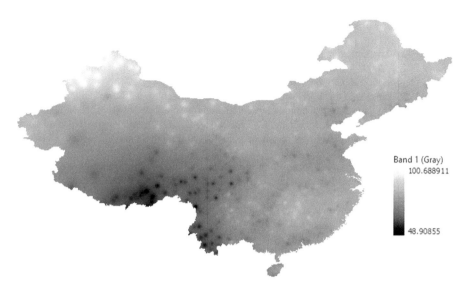

FIGURE 14.11 Standard deviation of annual mean solar radiation for China using IDW2.

However, what we have achieved is to create a continuous representation of solar insolation that covers the entirety of our study area from discrete points. There are also other interpolation methods, such as kriging, which could be used in subsequent GIS work. This chapter focused on environmental data, and while interpolation in socio-economic data is less readily employed, it can certainly be implemented. See Rastogi et al. (2020) for a recent example where they interpolated incomplete census information to inform crimes against women in India, incidentally, supporting SDG5.

14.3.1 Test Yourself

If you want to test yourself on the learning outcomes of this chapter, complete the following:

a. Repeat the IDW interpolation for standard deviation using a distance coefficient P of 2. Result presented in Figure 14.11 for reference.
b. Investigate using visual analysis and the raster calculator to identify if locations with a high mean and low standard deviation align.

REFERENCES

Joshi, S., Mittal, S., Holloway, P., Shukla, P.R., Ó Gallachóir, B. and Glynn, J., 2021. High resolution global spatiotemporal assessment of rooftop solar photovoltaics potential for renewable electricity generation. *Nature Communications*, 12(1), pp. 1–15.

Rastogi, A., Sridhar, S. and Gupta, R., 2020. Comparison of different spatial interpolation techniques to thematic mapping of socio-economic causes of crime against women. In *2020 Systems and Information Engineering Design Symposium (SIEDS)* (pp. 1–6). IEEE. https://ieeexplore.

ieee.org/abstract/document/9106690?casa_token=ziQw6CsJDmIAAAAA:-zlg82S1tooq1t8ROxBbuJKwVFnAm7E3jycb-mlI48CXVanH2s0LpKiJ34DQ26cpwA7fvCsmK#citations

Safriel, U., 2009. Deserts and desertification: Challenges but also opportunities. *Land Degradation & Development*, 20(4), pp. 353–366.

Tang, W.J., Yang, K., Qin, J., & Min, M. (2013). Development of a 50-year daily surface solar radiation dataset over China. Science China-earth Sciences, 56(9), 1555–1565.

Tang, W.J., Yang, K., Qin, J., Cheng, C.C.K., & He, J., (2010). Solar radiation trend across China in recent decades: a revisit with quality-controlled data. Atmospheric Chemistry and Physics, 11(1), 393–406.

Tang, W., 2019. *Daily Average Solar Radiation Dataset of 716 Weather Stations in China (1961–2010)*. National Tibetan Plateau Data Center, DOI: 10.11888/AtmosphericPhysics.tpe.249399.file.

Tobler, W.R., 1970. A computer movie simulating urban growth in the Detroit region. *Economic Geography*, 46(sup1), pp. 234–240.

World Food Programme (2019). *World Administrative Boundaries – Countries and Territories [dataset]*. Available from: https://public.opendatasoft.com/explore/dataset/world-administrative-boundaries/information/. Accessed March 1, 2022.

Section V

Spatial Analysis: Geoprocessing

15 Site Selection – Multiple Criteria Assessment

15.1 INTRODUCTION AND LEARNING OUTCOMES

In this chapter, we use multiple methods of geoprocessing spatial data, primarily vector data, including clip, intersect, union, and erase. We implement these tools to support the achievement of SDG6.1. This target aims to achieve universal and equitable access to safe and affordable drinking water. Here we use geoprocessing to identify a suitable site for a new water treatment plant through a multiple criteria assessment (MCA) in Uganda. The intervention of water, sanitation, and hygiene services in Uganda has been found to increase water quality in many rural schools, which has positive implications for health and cognitive development for school children (Morgan et al. 2021). Despite this, in some regions of the country, it has been estimated that less than 50% of households have safe drinking water (Lauer et al. 2018).

Wastewater treatment plants are essential to supporting SDG6, and even partial achievement of this goal would benefit humankind given the importance of clean and accessible water for sustainability. While the implementation of such plants is complex and nuanced, GIS can support informed decision making as to where best to install such infrastructure. For an interesting article on the potential for recycled wastewater to support SDG6, please see Tortajada (2020), while for a detailed discussion on GIS and MCA, please see Greene et al. (2011). In essence, MCA integrates and combines multiple datasets, which are then used to identify (or assess) areas that satisfy a pre-specified range of criteria (or multiple criteria). There are various selection criteria that could be used to support the selection of a site for a wastewater treatment plant (Massoud et al. 2009; Kanwal et al. 2020; Lizot et al. 2021). At a minimum, these five criteria are a recurring feature of most studies:

- Proximity to a major river
  - As close as possible
- Situated near the zones to be served
  - As close as possible to allow economic regrouping of the water sewers
- Access to the site by an existing road
  - As close as possible to avoid construction costs
- Presence of protected areas
  - Avoid zones to be protected
- A river or lake in close proximity
  - Avoid affecting minor rivers and lakes

We incorporate these five variables into our site selection process so that we can work through a manageable example; however, there are many other variables that

TABLE 15.1
List of Inclusion and Exclusion Variables

Inclusion	Exclusion
Near major rivers	Away from minor rivers
High population density	Away from protected areas
Near a road	

we could and perhaps should consider when working on such a case study. For our purposes it is sufficient to consider these five to demonstrate the GIS tools.

Firstly, it is useful to consider our variables as those that are desirable and subsequently included in our site selection, and those that are undesirable and excluded. We term these inclusion and exclusion variables. It is useful to tabulate these, so that we are clear exactly what analysis we need to implement for specific variables. The five variables are reported as inclusion and exclusion variables in Table 15.1.

By the end of this chapter, you will have completed three learning outcomes, and you should be able to:

- Implement data manipulation using geoprocessing tools, such as buffer, dissolve, and clip, to create new spatial information
- Undertake multiple criteria analysis using geoprocessing tools, such as intersect, union, and erase, to identify suitable and unsuitable areas
- Use a quantitative method to select a final suitable site for a new wastewater treatment plant

15.2 CASE STUDY: SDG6 LOCATING A NEW WASTEWATER TREATMENT PLANT IN UGANDA

1. Open a New Empty Project in QGIS.
2. Add all the layers from the Ch15_Data GeoPackage.
3. Right click on Uganda and zoom to the layer.

There are five layers present in this GeoPackage, shown in Figure 15.1. MajorRivers is a spatial layer from the World Bank (2018) licensed under CC-BY-4.0 of all major global rivers. Protected is a spatial layer from the Open Sustainability Institute (2017) licensed under CC0 1.0 of all protected areas in Uganda. Roads is a spatial layer from CIESIN (2013) of roads within Uganda. Uganda is an outline of the country from the world-administrative-boundaries layer from the World Food Programme (2019) licensed under an Open Government License v3.0. MinorRivers is a spatial layer from the Uganda Bureau of Statistics (2012) of the rivers and lakes in the country, published in the public domain.

The first thing we must do is ensure that all the layers are projected in the same coordinate system. As we are working in Uganda and because our geoprocessing tools require linear distances, we want to project everything into UTM Zone 36S.

Site Selection – Multiple Criteria Assessment 253

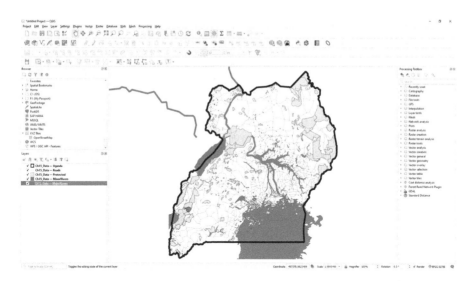

FIGURE 15.1 Screenshot of the five layers within the Ch15 Data folder, including the outline of Uganda, major and minor rivers, roads, and protected areas.

This is EPSG code 32736. To reinforce the fundamental GIS skills, one layer is not projected in this CRS.

4. Using the information tab in Properties, identify the layer that is not projected using this CRS. (Hint: this needs to be completed for each layer until it is found.)
5. Reproject the layer to UTM 36S (hint Vector > Data Management Tools > Reproject Layer).

The layer in question was MinorRivers, and hopefully it is now saved in the GeoPackage as MinorRivers_36S. It would be a good idea to remove the original MinorRivers layer to avoid confusion.

Now we are ready to start geoprocessing. MajorRivers spans the entire continent (i.e., the whole of Africa), while the other layers are all captured at a national scale (i.e., Uganda). The first step in this workflow is to clip MajorRivers to the country level. We have undertaken several clips so far, meaning we should know that this tool applies a cookie cutter procedure to the layer in question.

6. Navigate through the tab Vector > Geoprocessing Tools > Clip.
7. Specify MajorRivers as the Input Layer, Uganda as the Overlay Layer, and save the resultant layer as MajorRivers_Uganda in the GeoPackage.
8. Click Run.

Now, four of our variables (roads, major rivers, minor rivers, and protected areas) need to be assessed based on distance. To achieve this, we are going to use the buffer tool. Again, we introduced buffering in Chapter 5, before refining our skills in Chapter 11. Therefore, we know that this creates a zone of proximity around our features, which we can then use as our criteria in the site selection process. This is important as distance to these features is an integral part of the analysis, but also to overlay layers to identify a suitable site we need all layers to be in polygon format.

9. Navigate through the tab Vector > Geoprocessing Tools > Buffer.
10. Choose Protected as the Input layer, the buffer distance as 500 m, ensure Dissolve result is ticked, and choose an informative name (i.e., Protected_buffer500) to save in the GeoPackage.

Remember in Chapter 11 we discussed the importance of dissolving our buffers to ensure data redundancy is minimal and speed up processing when combining the outputs in geoprocessing operations. This is such an instance, so be sure to dissolve all the buffers.

11. Repeat the buffering process for the following layers, using the specified distances:
    - Major Rivers (the version clipped to Uganda) – 5000 m
    - Minor Rivers (the version reprojected into UTM 36S) – 500 m
    - Roads – 500 m

Investigate the minor and major rivers by turning off the other layers and visualizing only these. We can see that they contain some of the same rivers. Therefore, to allow the analysis to be implemented and allow differentiation between the two types, we must erase the major rivers from the minor rivers, so that the minor rivers only contain the minor rivers. To do this we use the tool Difference, which is synonymous with the GIS term/function Erase.

12. Navigate through the tab Vector > Geoprocessing Tools > Difference.
13. Select MinorRivers_Buffer500 as the Input Layer.
14. Select MajorRivers_Buffer5000 as the Overlay.
15. Specify the new layer as Minor_NotMajor in the GeoPackage.
16. Ensure your dialogue box resembles Figure 15.2. Click Run.

The new layer is now representative of only the minor rivers, and the difference between all rivers and minor rivers only is shown in Figure 15.3.

If we return to Table 15.1, we have now completed four of the five requirements, with only high population density outstanding; however, this layer is currently in raster format. Firstly, we need to add this layer to the project and then convert it.

17. Add the raster layer UGA_ppp_v2b_2015_UNadj.tif from the extracted folder.

This layer represents the population density of Uganda from WorldPop (2013) licensed under CC-BY-4.0.

Site Selection – Multiple Criteria Assessment

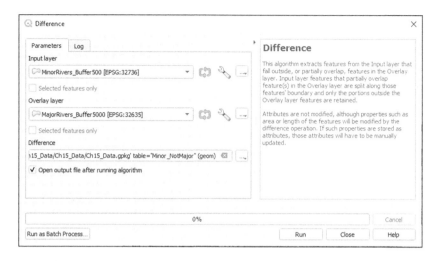

FIGURE 15.2 Screenshot of the parameters for the difference tool.

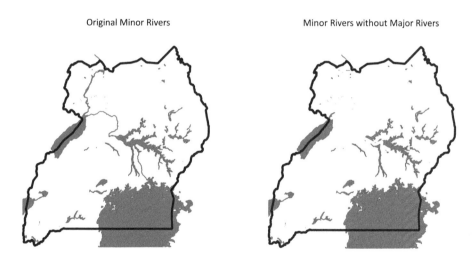

FIGURE 15.3 Difference between original minor rivers layer (left) and after major rivers has been erased using the difference tool (right).

18. Navigate through the tab Raster > Conversion > Polygonize (Raster to Vector).
19. Choose UGA_ppp_v2b_2015_UNadj as the Input Layer.
20. Set Band 1 as the Band Number.
21. Keep the name of the new field as DN.
22. Save the output in your working directory as a shapefile (there is no option to save this in the GeoPackage). Use PopDen as the name.
23. Ensure your dialogue box resembles Figure 15.4. Click Run.

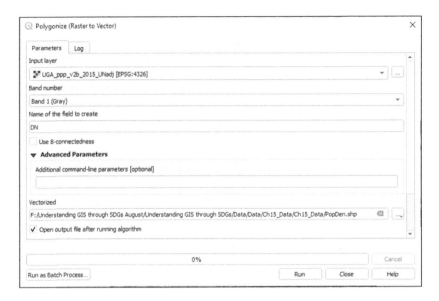

FIGURE 15.4 Screenshot of parameters needed for the polygonize tool.

We now have a polygon representation of the population density raster file. As we know from Chapter 6, when we convert from rasters to polygons, we need to check the geometry associated with the new polygon.

24. In the Processing Toolbox, navigate to Vector Geometry > Check Validity.
25. Select the new PopDen layer as the input and specify GEOS as the method.

If this function reports invalid geometries, fix them using the methods outlined in Chapter 6 (and below).

26. Click on the Processing Toolbox and search for Fix Geometries.
27. Choose PopDen as the input, save the output as PopDen_Fix.
28. Click Run.

Returning to the site selection, we need locations that have a high population density, but without a reference as to what constitutes average population density, we may be unsure as to the best way to proceed. We can view the statistics associated with the population density field.

29. Navigate through the tab Vector > Analysis Tools > Basic Statistics for Fields.
30. Select PopDen is the Input layer.
31. Select DN as the field.
32. Keep the temporary name as we do not need this saved permanently.
33. Ensure your dialogue box resembles Figure 15.5. Click Run.

The results viewer should pop up (if not Processing > Results Viewer).

Site Selection – Multiple Criteria Assessment

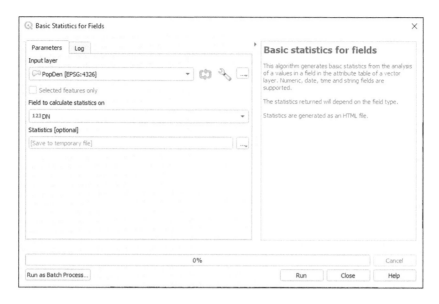

FIGURE 15.5 Screenshot of parameters needed for the basic statistics for fields tool.

34. Click on the html link.

Use this to identify the mean value. We know that the mean value represents the central tendency (or middle point), and as such, this would be a good quantitative value to base our value for 'high' population density. The statistics should resemble Figure 15.6.

35. Return to QGIS.
36. Open Select Features by Value tool.
37. Specify a query that selects features that have a DN greater than or equal to the mean (which we should know is 17.21).
38. Click Select Features.

There should be 87,560 features selected. Next, we want to make this selection permanent.

39. Right click on the layer in the Layers Panel.
40. Export > Save Selected Features As.
41. Save the layer as HighPopDen in the GeoPackage.
42. Select UTM Zone 36S as the CRS.

This would also be a good time to dissolve this new layer as we are not interested in the unique values associated with features.

43. Navigate through the tab Vector > Geoprocessing > Dissolve.
44. Select either HighPopDen as the input.

FIGURE 15.6 Screenshot of the results for the basic statistics for fields tool.

TABLE 15.2
List of Inclusion and Exclusion Variables and Updated Layer Names

Inclusion	Exclusion
Near major rivers	Away from minor rivers
High population density	Away from protected areas
Near a road	
New Names	
MajorRivers_Buffer5000m	Minor_NotMajor
HighPopDen_Dissolve	Protected_Buffer500m
Roads_Buffer500m	

45. Save the layer in the GeoPackage as HighPopDen_Dissolve.
46. Click Run.

We have all five inclusion and exclusion variables, now we need to combine them. The new layer names are reported in Table 15.2 if you have been following along with the naming convention.

There are two types of geoprocessing overlay tools we use here: union and intersect. Union combines features from two polygon layers at a time to create a new

Site Selection – Multiple Criteria Assessment

polygon layer. The new layer contains all polygons and all attributes from both layers. It is the spatial equivalent of a Boolean OR statement. Intersect combines features and attributes from two layers into a new layer that retains the geographic area common to both input layers. It is the spatial equivalent of a Boolean AND statement. Whenever we perform a union or intersect, all attributes are also joined, meaning we can perform queries across both layers, and make new associations between attributes and geometry (which we explore in more detail in the next chapter).

47. Navigate through the tab Vector > Geoprocessing Tools > Union.
48. Select Minor_NotMajor as the input layer.
49. Select Protected_buff500 as the overlay layer (in practice it does not matter what order we undertake this).
50. Save the output in the GeoPackage. For this operation, we are only overlaying two layers, but we cannot overlay more than two, meaning we might want to name our permanent layer something informative, such as Union_minor_protected.

It is important to note here as well that we cannot combine layers that are not polygons. Therefore, we must ensure these are the layers we have buffered, otherwise the function will fail, or worse it will run but provide us with incorrect data.

51. Ensure your dialogue box resembles Figure 15.7. Click Run.

If we zoom to this layer, we see that the union has created several smaller polygons. Therefore, a dissolve would be a good idea to remove data redundancy.

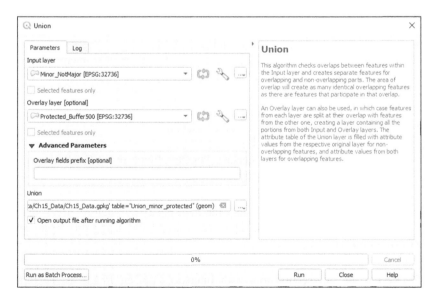

FIGURE 15.7   Screenshot of parameters for the union tool.

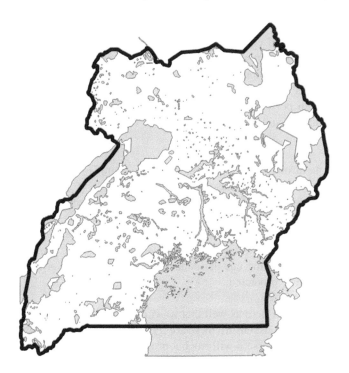

FIGURE 15.8 Screenshot of the dissolved union layer, which includes both protected areas and minor rivers.

52. Dissolve the new union layer using the geoprocessing tool Dissolve (i.e., Union_minor_protected) and name it Union_dissolve.

Your map should resemble that shown in Figure 15.8.

Now we want to combine the inclusion layers. As we are only interested in areas where ALL of these are found, we use intersect.

53. Navigate through the tab Vector > Geoprocessing Tools > Intersection.
54. Select HighPopDen_Dissolve as the input layer.
55. Select Roads_buffer500 as the overlay layer.
56. Save the resultant layer as Intersect_PopDen_Roads in the GeoPackage.
57. Ensure your dialogue box resembles Figure 15.9. Click Run.

This returns a spatial layer where only the high-population-density areas and road buffers are found. Next, we must repeat the intersection, using this layer and the major rivers buffer.

58. Repeat the intersection tool with Intersect_PopDen_Roads as the input layer, MajorRivers_buffer5000 as the overlay, and name the final layer as Intersect_Major_Roads_PopDen in the GeoPackage.

Site Selection – Multiple Criteria Assessment

FIGURE 15.9   Screenshot of parameters for the intersect tool.

59. Dissolve the Intersect_Major_Road_PopDen layer and save it in the GeoPackage as Intersect_dissolve.

Your map should resemble Figure 15.10. Please note to ensure visibility I have increased the stroke width to 1 on the Intersect_Dissolve layer so it is visible.

We now have two layers. The union layer represents all locations where we cannot locate a new wastewater treatment plant, while the intersect layer represents all locations where we would ideally locate one. We can immediately see the difference between them in terms of spatial coverage. The final step is to erase the exclusion layer (union) from the inclusion layers (intersect).

60. Navigate through the tab Vector > Geoprocessing Tools > Difference.
61. Select the Intersect_Dissolved as the Input layer.
62. Select the Union_Dissolved as the Overlay layer.
63. Save the resultant layer as Final_Sites in the GeoPackage.

It is key here that we have dissolved the union and intersect. If not, you could be sitting here waiting for this to run for a couple of hours as opposed to a couple of seconds…

64. Ensure your dialogue box resembles Figure 15.11. Click Run.

Turn all the layers off except for the Final_Sites. There should be at least eight sites that could be a suitable location for a new wastewater treatment plant, presented in Figure 15.12.

FIGURE 15.10 Screenshot of the dissolved intersect layer, which includes major rivers, high population density, and roads.

FIGURE 15.11 Screenshot of parameters needed for final overlay using the difference tool.

FIGURE 15.12 Screenshot of the difference (erase) layer, which returns only suitable sites for a new wastewater treatment plant.

We have identified several locations that satisfy our criteria. However, in a lot of site selection studies, people report several locations, but do not select a final location. Here we use a spatial layer of borehole locations from the UNHCR WASH (2022) program.

65. Navigate to the UNHCR WASH (2022) geoportal: https://wash.unhcr.org/wash-gis-portal/.
66. Click on the Download option, ensure Uganda is ticked, and then click on the Download Selected Countries button.
67. Choose *.csv as the format.

This will download to your device.

68. Add the layer Boreholes by navigating through the tab Layer > Add Layer > Add Delimited Text Layer. The coordinate system is latitude and longitude using a CRS of WGS 84 (EPSG 4326).
69. Ensure Boreholes is activated in the Layers Panel and navigate to Select by Expression.
70. Specify a query to select a borehole that does not fall into suitable pH levels (suitable levels between 6.5 and 8.5), the expression is provided in Figure 15.13. For reference there should be 24 boreholes that do not fall within this range.

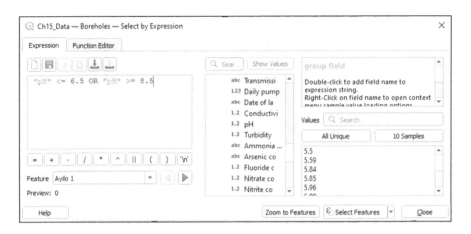

FIGURE 15.13   Select by expression query to return unsuitable pH levels.

71. Save the selected features as a permanent layer.
72. Use the Measure Line tool to quantify the distance between suitable sites and a borehole (i.e., well) that has an undrinkable pH level.

15.3   CASE STUDY CONCLUDING REMARKS

For reference, there is not a suitable site for almost 60 km. This site is in the center of the country and the borehole at Bwcyale, and the site south-west of Lira in Figure 15.14. There are quite a lot of areas with undrinkable water, meaning that perhaps efforts need to be made to refine our variables, and choose a different set of selectors that might have the greatest impact on the most people. This is where an alternative site selection method might be optimal, using a set of weightings. Similarly, if we were to change the buffer from the major rivers from 5 to 20 km, this would give us a suitable site closer to an existing borehole. This last point emphasizes the importance of user decisions in implementing this type of analysis and is another reason why the consideration of scale and neighborhoods is fundamental to good GIS practice.

This chapter has introduced us to the concept of MCA, whereby we use geoprocessing to manipulate and transform our spatial data to match a set of criteria. We have reinforced earlier skills such as queries, clips, and buffers, as well as introducing new geoprocessing tools such as union, intersection, and difference. These latter tools are the spatial equivalent of using Boolean logic, and we revisit this in the subsequent chapter. Finally, it's important when undertaking this type of MCA to explore the possible final locations. Oftentimes the final step of choosing an 'optimal' location can be overlooked. Therefore, using the measure tool to locate the closest site is a quantitative method that can add significant value to such work or highlight limitations to the process and suggest refinements, as it has in this instance. We revisit many of these geoprocessing tools in the next chapter to reinforce the learning outcomes associated with them.

Site Selection – Multiple Criteria Assessment

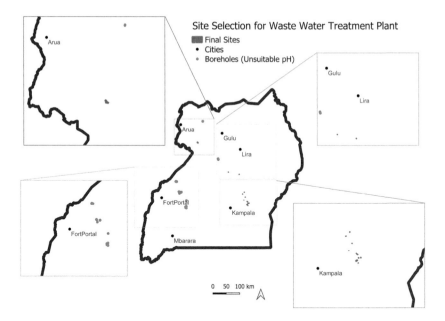

FIGURE 15.14 Map of final suitable sites using the initial selection criteria, alongside boreholes with unsuitable pH and major cities in Uganda for reference.

15.3.1 Test Yourself

If you want to test yourself on the learning outcomes of this chapter, complete the following:

a. Re-run the buffer for major rivers at 20 km.
b. Re-run the second intersect and erase the union from this output.
c. Identify a suitable site approximately 3 km away from a borehole.

REFERENCES

Center for International Earth Science Information Network – CIESIN – Columbia University, and Information Technology Outreach Services – ITOS – University of Georgia (2013). *Global Roads Open Access Data Set, Version 1 (gROADSv1)*. NASA Socioeconomic Data and Applications Center (SEDAC), Palisades, NY. https://doi.org/10.7927/H4VD6WCT. Accessed March 1, 2022.

Greene, R., Devillers, R., Luther, J.E. and Eddy, B.G., 2011. GIS-based multiple-criteria decision analysis. *Geography Compass*, 5(6), pp. 412–432.

Kanwal, S., Sajjad, M., Gabriel, H.F. and Hussain, E., 2020. Towards sustainable wastewater management: A spatial multi-criteria framework to site the Land-FILTER system in a complex urban environment. *Journal of Cleaner Production*, 266, p. 121987.

Lauer, J.M., Duggan, C.P., Ausman, L.M., Griffiths, J.K., Webb, P., Bashaasha, B., Agaba, E., Turyashemererwa, F.M. and Ghosh, S., 2018. Unsafe drinking water is associated with environmental enteric dysfunction and poor growth outcomes in young children in rural southwestern Uganda. *The American Journal of Tropical Medicine and Hygiene*, 99(6), p. 1606.

Lizot, M., Goffi, A.S., Thesari, S.S., Trojan, F., Afonso, P.S. and Ferreira, P.F., 2021. Multi-criteria methodology for selection of wastewater treatment systems with economic, social, technical and environmental aspects. *Environment, Development and Sustainability*, 23(7), pp. 9827–9851.

Massoud, M.A., Tarhini, A. and Nasr, J.A., 2009. Decentralized approaches to wastewater treatment and management: Applicability in developing countries. *Journal of Environmental Management*, 90(1), pp. 652–659.

Morgan, C.E., Bowling, J.M., Bartram, J. and Kayser, G.L., 2021. Attributes of drinking water, sanitation, and hygiene associated with microbiological water quality of stored drinking water in rural schools in Mozambique and Uganda. *International Journal of Hygiene and Environmental Health*, 236, p. 113804.

Open Sustainability Institute (2017). *Uganda Protected Areas*.

Tortajada, C., 2020. Contributions of recycled wastewater to clean water and sanitation sustainable development goals. *NPJ Clean Water*, 3(1), pp. 1–6.

Uganda Bureau of Statistics (2012). *Lakes and Rivers: Uganda, 2005*. Uganda Bureau of Statistics. Available from: http://purl.stanford.edu/fh022bz4757.

World Bank (2018). *Major Rivers of the World*. Available from: https://datacatalog.worldbank.org/search/dataset/0042032.

World Food Programme (2019). *World Administrative Boundaries – Countries and Territories [dataset]*. Available from: https://public.opendatasoft.com/explore/dataset/world-administrative-boundaries/information/. Accessed March 1, 2022.

16 Risk Analysis – Unique Condition Unit

16.1 INTRODUCTION AND LEARNING OUTCOMES

In this chapter, we reinforce the geoprocessing methods introduced in previous chapters, as well as implementing some new raster analytical techniques. We implement these tools to support the achievement of SDG11.5. This target aims to significantly reduce the number of deaths and people affected by disasters, as well as substantially decrease the direct economic losses relative to global gross domestic product (GDP). To achieve this, we complete a risk analysis of landslides in Arequipa, Peru (Figure 16.1).

Peru is perhaps the country where the most infamous landslide happened in 1970, with an approximate death toll of up to 70,000 people. While this is a stark example of the risk that landslides pose, there have been multiple landslides since then that continue to cause deaths and displace large numbers of people. GIS can be used to establish landslide risk models, which provide an early warning system that supports resilience of local communities and can significantly reduce death rates. Landslide susceptibility is impacted by several factors, with the more serious often triggered by earthquakes which are difficult to predict; however, most are caused by a combination of three factors:

- Slope
- Soil Type
- Flow of Water

There are several methods of predicting or quantifying landslide susceptibility in GIS (Huabin et al. 2005; Akgun et al. 2012; Lee 2019). In this chapter, we use a method called the unique condition unit. This method applies the concept that if a landslide has occurred in a particular set of conditions previously and if those conditions occur elsewhere, the other locations with the same conditions are also susceptible. This approach is primarily deductive in nature, where we describe the relationships between event occurrences and a set of geospatial predictors. We identify the range of environmental values for which landslide events have occurred in, before combining them spatially.

By the end of this chapter, you will have completed four learning outcomes, and you should be able to:

- Competently undertake geoprocessing tools such as clip, intersection, union, and difference

FIGURE 16.1 Arequipa, Peru. Steeped terraces in the foreground, with the volcano Misti in the background. (Photo credit: Paul Holloway.)

- Perform slope analysis using elevation data
- Implement multiple spatial joins
- Perform advanced queries to generate new data in the attribute table

16.2 CASE STUDY: SDG11.5 LANDSLIDE RISK IN AREQUIPA, PERU

1. Open a New Empty Project in QGIS.
2. Add the four vector files and one raster file from the extracted folder.

Your map should resemble Figure 16.2. Global_Landslides is a point file of global landslide locations from Kirschbaum et al. (2010, 2015) downloaded via Data.gov licensed under Open Data Commons Open Database License (ODbL) v1.0. SoilType_Fix is a polygon representation of soil type converted from the WISE30sec soil properties global grid (Batjes 2016) licensed under CC-BY-3.0. This is the file we fixed the geometries for in Chapter 6. PPT is a polygon representation of the average annual precipitation grouped into categories of 500 mm over a total year converted from Bioclim 12 into a polygon format (Hijmans et al. 2005), licensed under Creative Commons Attribution-Sharealike 4.0 International License. Peru_Elevation is from the Global Bathymetry and Elevation Digital Elevation Model: SRTM30_PLUS v8 (Becker et al. 2009) licensed under CC-BY-3.0. Finally, Arequipa is a polygon representation of the

Risk Analysis – Unique Condition Unit

FIGURE 16.2 Screenshot of the five layers for use in the chapter in the QGIS interface.

administrative unit for the region Arequipa digitized by the author. The three geospatial predictors (soil, ppt, and elevation) are all clipped to the extent of Peru, while the landslide layer represents locations at a global extent. Therefore, the first job we need to undertake is to clip all layers to the Arequipa region.

3. Navigate through the tab Vector > Geoprocessing Tools > Clip.
4. Select Global Landslides as the Input Layer.
5. Select Arequipa as the Overlay Layer.
6. Save the resultant layer in the GeoPackage as Arequipa_Landslides.
7. Click Run.

The landslide points should now be 'clipped' to the extent of Arequipa, with five landslides recorded there during the time period of the dataset.

8. Repeat the clip for SoilType_Fix and PPT. Save the layers using a suitable name, such as adding Arequipa as a prefix.
9. Navigate through the tab Raster > Extraction > Clip Raster by Mask Layer. Extract the Peru_Elevation raster to the extent of Arequipa. Again, save the layer using a suitable name as a *.tif. We also need to specify the NoData value as −9999, otherwise this will return 0 as the value for NoData which has implications for our analysis down the pipeline.

Next, we need to join the different environmental variables to the landslide point layer so that we can assess the unique conditions. The first layer we join to the landslide points is Arequipa_SoilType. However, before we get started on this layer, it

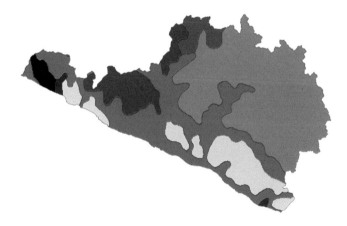

FIGURE 16.3 Soil type in Arequipa.

would be useful to visualize the different layers so that we can begin to intuitively explore the differences.

10. Right click on Arequipa_SoilType layer to open Properties.
11. Open the Symbology tab.
12. Change the option from Single Symbol to Categorized.
13. Select Soil_Type_ as the Value.
14. Choose a color palette, such as Inferno.
15. Click Classify.
16. Click OK.

Your map should resemble Figure 16.3. We should see that there are several coarse categories of soil type across the region. The soil codes are provided in Batjes (2016). If we turn on the Arequipa_Landslide layer, we can see that the landslides occurred in only a small number of these categories. To quantify this, we want to join the polygon values to the points. While we have explored table joins extensively in the book thus far, in this chapter we implement spatial joins. This requires a different tool that works in a similar way to the tool we used in Chapter 3, but instead of sampling a raster layer, we sample a polygon layer.

17. In the Processing Toolbox, navigate to Join Attributes by Location, which is in Vector General and open this tool.
18. Select Arequipa_Landslides as the Base Layer.
19. Select Intersects.
20. Select Arequipa_SoilType as the Join Layer.
21. Click on options … for fields to add, and make sure that Soil_Type_ is ticked. Click OK.
22. Save the joined layer to the GeoPackage. As we are generating three joins, name this Arequipa_Landslides_Soil so it is clear which variables have already been joined.

Risk Analysis – Unique Condition Unit

FIGURE 16.4 Screenshot of the parameters needed for the join attributes by location tool.

23. Ensure your dialogue box resembles Figure 16.4. Click Run.

We should check that the join has worked.

24. Open the attribute table of the Arequipa_Landslides_Soil layer and scroll across to the right-hand column to see if the soil has appended.

The final column should read Soil_Type_. If it does not, check to make sure the parameters are correct. Now we have joined these two layers, the next step is to identify all soil types that have been associated with a landslide. We could do this visually as there are only five points, but for larger layers we need an automated tool.

25. Navigate through the tab Vector > Analysis Tools > List Unique Values.
26. Choose Arequipa_Landslides_Soils as the Input Layer.
27. In Target Field, click on … and make sure that Soil_Type_ is the only layer ticked. Click OK.

28. Save the unique values as a layer in the GeoPackage called 'Soil_Type_Values'.
29. Ensure your dialogue box resembles Figure 16.5. Click Run.

This tool generates both a table (which should have loaded in the Layers Panel) and a html file. We want to open the table.

30. Right click on Soil_Type_Values table in the Layers Panel, and open the attribute table.

The table lists two unique soil types, as shown in Figure 16.6. Now that we have this information, we can create a new layer representing soil types, but that only consists of these layers.

31. Navigate to Select by Expression, activating Arequipa_SoilType by clicking on it in the Layers Panel.

FIGURE 16.5 Screenshot of parameters for the list unique values tool.

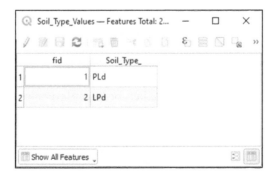

FIGURE 16.6 Screenshot of attribute table for unique values.

Risk Analysis – Unique Condition Unit

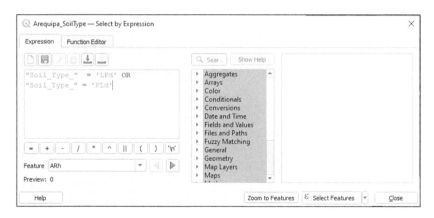

FIGURE 16.7 Screenshot of select by expression.

32. Type the following expression: "Soil_Type_" = 'LPd' OR "Soil_Type_" = 'PLd'.
33. Ensure your dialogue box resembles Figure 16.7. Click Select Features (for reference there are two features selected).
34. Right Click on the Arequipa_SoilType layer > Export > Save Selected Features As.
35. Save this in the GeoPackage as 'Soil_Risk'.

The categories of soil are quite broad, so this layer represents quite a large proportion of the region; however, this process has removed approximately half of the soil types in the area.

Next, we explore the impact of rainfall. Here we use a variable that represents the average total yearly precipitation, which is bioclimatic variable 12 from Hijmans et al. (2005). Again, this is most likely a simplification of 'water' when considering landslides, especially when we consult the attribute table of landslide causes including 'rain' and 'downpour'. For example, the risk of a landslide event would most likely be higher in some areas if 200 mm of rainfall fell in 1 hour as opposed to 1 month. Therefore, when working on such analysis, care needs to be taken in identifying relevant variables to use. Again, it is always useful to visualize our data using symbology. Use the following steps to change the symbology for the PPT layer for the whole of Peru:

36. Open the properties of the PPT layer for the country extent.
37. Change the symbology from Single Symbol to Categorized.
38. Set the value as PPT.
39. Change the color palette to blues.
40. Click Classify.
41. Click Apply.

Your map should resemble Figure 16.8. Here we can see the annual precipitation grouped into 500 mm categories, with a low rainfall along the coast and a clear

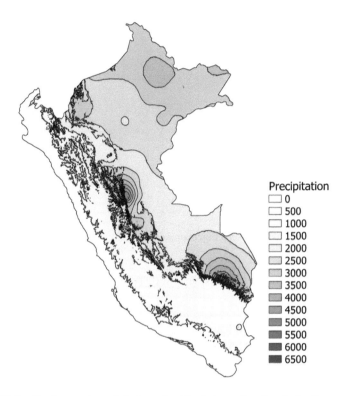

FIGURE 16.8 Total annual precipitation visualized at a national scale for Peru.

increase in rain as the orthographic rainfall effect occurs east of the Andes. In the Arequipa region, we can see that there is generally low rainfall, with what appears to be only two categories; however, for the process of rigor, we should quantify this using the methods identified.

As we progress through this book, it is useful to try and test yourself by undertaking steps without the instructions. The information is provided below, but it is the same as the functions we undertook for soil type. Therefore, try and attempt this on your own if possible (up to step 60). The only differences are the fact that we want to join this layer to the point layer Arequipa_Landslides_Soil, so that all attributes are joined in the same layer. While we visualized PPT at the national extent, keep the processing to the Arequipa region.

42. In the Processing Toolbox, search for Join Attributes by Location, and open this tool. Note there is a similar tool called Join Attributes by Location (summary), we do not want to use this tool.
43. Select Arequipa_Landslides_Soil as the Base Layer.
44. Select Intersects.
45. Select Arequipa_PPT as the Join Layer.
46. Click on options ... for fields to add, and make sure that PPT is ticked, and click OK.

Risk Analysis – Unique Condition Unit

47. Save the joined layer in the GeoPackage. Again, name this Arequipa_Landslides_Soil_PPT to remind us which variables have been joined to the landslides layer.
48. Click Run.
49. Open the attribute table to check this has joined correctly.

There should now be two attributes at the end of the table, Soil_Type_ and PPT.

50. Navigate through the tab Vector > Analysis Tools > List Unique Values.
51. Choose Arequipa_Landslides_Soils_PPT as the Input Layer.
52. Click on … and make sure that PPT is the only layer ticked.
53. Save the unique layer as 'PPT_Table'.
54. Click Run.
55. Right click on PPT_Table in the Layers Panel and open the attribute table.

There should be two values, 500 and 1000. Now that we have this information, we can again create a new layer representing ppt, but that only consists of these values.

56. Navigate to Select by Expression, ensuring that Arequipa_PPT is activated in the Layers Panel.

As these are numeric values, and all ppt values under 1000 mm per year are returned, we can use a numeric operator to return our values.

57. Type the following expression: "PPT" <= 1000.
58. Click Select Features (there should be 81 selected features).
59. Right Click on the Arequipa_PPT layer > Export > Save Selected Features As.
60. Save this in the GeoPackage as 'PPT_Risk'.

As hinted at earlier, there is very little variation in annual precipitation in the region, meaning that the whole of the region has been returned as high risk for PPT. Finally, we are now interested in identifying the slope angle in locations where landslides have occurred.

61. Navigate to the Processing Toolbox and locate Slope. This should be in the Raster Terrain Analysis toolbar. Open the tool.
62. Select the Arequipa_Elevation as the input layer.
63. Change the z factor to 0.00000912.

This is a vertical exaggeration, which can be useful when the Z units differ from the X and Y units (e.g., feet and meters). In our example, our X and Y coordinates are in angular degrees, but our Z values (i.e., elevation) are in meters. Therefore, we need to scale the two. As Peru spans approximately 20° of latitude, we will use a z factor score of 0.00000912. We can note here that z values can range from 0.00000898 at the equator to 0.00005156 at the poles. The alternative would be to re-project the elevation data into a planar coordinate system, because if our X, Y, and Z values are calculated using meters, then we can keep the scale to 1.

64. Save the output to your working directory as Arequipa_Slope.
65. Ensure your dialogue box resembles Figure 16.9. Click Run.

Your map should resemble Figure 16.10. We should pause here to note that slope is a scale-dependent variable, meaning such results will depend on the resolution of the input data (e.g., 30 m resolution v 300 m resolution). See Goodchild (2011) for a discussion. Below we have a detailed slope map, ranging from 0° to approximately 24°. If the maximum value is 60°, it is recording the NoData values outside of Arequipa as 0, and subsequently calculating slope based on this. This is most likely because −9999 was

FIGURE 16.9   Screenshot of the parameters for the slope tool.

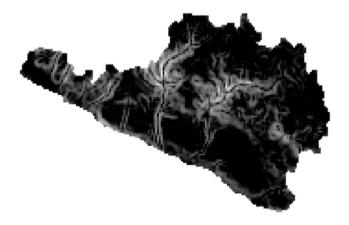

FIGURE 16.10   Screenshot of slope in the Arequipa region.

not specified in Step 9. If this is the case, return to Step 9 and re-mask the elevation data specifying the NoData value.

As the rest of our data is in vector format, we first group the values into categories and then convert the data. Aggregating the raw values into groups does lose information, and in the following chapters we explore methods of using raster data to analyze such phenomena so that we do not lose this information. However, to demonstrate tools we can use with vector data, we implement this now.

66. Navigate to Reclassify by table in the Processing toolbox, and open the tool.
67. Choose Arequipa_Slope as the input.
68. Set Band 1 as the band.
69. Click on … for the Reclassification table.

We want to specify a table that groups the values into 2.5° bins.

70. Complete the table using 2.5° bins, as shown in Figure 16.11, and then click OK to close the table. Do not click Run, as otherwise this will try to complete the tool with parameters still missing (and most likely bug out).
71. Save the output layer as Arequipa_Slope_Reclass in your working directory.
72. Keep the rest of the options as default.
73. Click Run.

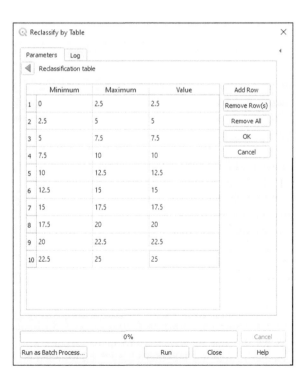

FIGURE 16.11 Screenshot of the reclassify by table parameters.

Next, we want to polygonize this data.

74. Navigate through the tab Raster > Conversion > Polygonize.
75. Set Arequipa_Slope_Reclass as the input.
76. Set Band 1 as the band number.
77. Set the Name of the Field to Create as 'slope'.
78. Save the layer as a shapefile in the working directory called Arequipa_Slope_Polygon (note there is no option to save in a GeoPackage, only as a GeoPackage, so to avoid overwriting all our data save as a shapefile).
79. Keep the rest of the settings as default.
80. Click Run.

Next, we want to join the slope polygon layer to our ever-growing landslide point file (the one we now have attribute information for soil and ppt). Again, try to complete these steps independently up to Step 98.

81. In the Processing Toolbox, search for Join Attributes by Location, and open this tool.
82. Select Arequipa_Landslides_Soil_PPT as the Base Layer.
83. Select Intersects.
84. Select Arequipa_Slope_Polygon as the Join Layer.
85. Click on options ... for fields to add, and make sure that Slope is ticked.
86. Save the joined layer in the GeoPackage as Arequipa_Landslides_Soil_PPT_Slope.
87. Click Run.

A warning will appear stating that this layer does not have a spatial index. As explained in Chapter 6, we can fix this quite simply; however, as we only have five points this will not slow down our processing. We can use the tool Create spatial index if we were working with larger files.

88. Open the attribute table to check this has joined correctly. There should now be three attributes in the table, Soil_Type_, PPT, and slope.
89. Navigate through the tab Vector > Analysis Tools > List Unique Values.
90. Choose Arequipa_Landslides_Soils_PPT_Slope as the Input Layer.
91. Click on ... and make sure that Slope is the only layer ticked.
92. Save the unique layer as 'Slope_Table'.
93. Click Run.
94. Right click on Slope_Table table in the Layers Panel, and open the attribute table.

There are four unique values, as shown in Figure 16.12: 5, 8, 10, and 13. This might seem unusual as we did not specify 8 or 13 in the reclassification function, but because of the integer format needed, these have been rounded up. We may want to consider cleaning this up, but for now we know that 8 means 5–7.5° slope angle, so we can push on.

Risk Analysis – Unique Condition Unit

FIGURE 16.12 Attribute table of unique slope values.

95. Navigate to Select by Expression (ensuring Arequipa_Slope_Polygon is selected in the Layers Panel) and type the following expression: "Slope" = 5 OR "Slope" = 8 OR "Slope" = 10 OR "Slope" = 13.
96. Click Select Features (there should be 1283 features selected).
97. Right Click on the Arequipa_Slope_Polygon layer > Export > Save Selected Features As.
98. Save this in the GeoPackage as 'Slope_Risk'.

We now have three risk layers for each of the environmental variables, Soil_Risk, PPT_Risk, and Slope_Risk, shown in Figure 16.13. We can turn all the other layers off.

To identify areas of high landslide susceptibility, we want to identify locations that contain features associated with Soil_Risk AND PPT_Risk AND Slope_Risk. To do this, we use the Intersection tool to return only locations that match all of these criteria.

99. Navigate through the tab Vector > Geoprocessing > Intersection.

We can only overlay two layers at once, so first we intersect Soil_Risk and PPT_Risk.

100. Choose Soil_Risk as the Input Layer.
101. Choose PPT_Risk as the Overlay Layer.
102. Save the output in the GeoPackage as Intersect_Soil_PPT.
103. Click Run.

This should have returned quite a large area where we have soil and ppt susceptibility. Next, we repeat this, using the newly created intersect layer and slope.

104. Open the Intersection tool again.

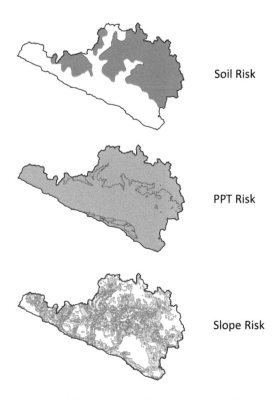

FIGURE 16.13 Screenshot of the three new risk layers that have been created.

105. Choose Intersect_Soil_PPT as the Input layer.
106. Choose Slope_Risk as the Overlay Layer.
107. Save the output in your working directory as Intersection_Soil_PPT_Slope.
108. Click Run.

Finally, we are going to dissolve this layer for presentation purposes.

109. Navigate through the tab Vector > Geoprocessing > Dissolve.
110. Set the input layer as Intersection_Soil_PPT_Slope.
111. We are not dissolving based on a field, so leave this as default.
112. Save the output as Susceptible in the GeoPackage.
113. Click Run.

Our results, presented in Figure 16.14, tell us that a large area of Arequipa, particularly the north-east, is susceptible to landslides based on our simple criteria. We could then subsequently use this layer to identify people and infrastructure at risk. One limitation of our work is that we have considered any combination of soil, precipitation, and slope, when in fact all the existing landslides occurred across unique combinations of these three variables. We can refine our approach to accommodate all these components. Using vector data, we need all three of our

Risk Analysis – Unique Condition Unit

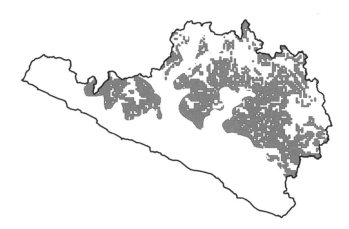

FIGURE 16.14 Results of the risk analysis that identify areas of landslide susceptibility in Arequipa, Peru, based on the three environmental criteria.

layers combined, so that the attributes can be compared. To do this, we implement a union.

114. Navigate through the tab Vector > Geoprocessing > Union.
115. Select Arequipa_SoilType as the Input layer.
116. Select Arequipa_PPT as the Overlay layer.
117. Save the output in the GeoPackage as Union_Soil_PPT.
118. Click Run.

The two layers are now overlaid with new polygons where we have different boundaries. Repeat this for Slope.

119. Navigate through the tab Vector > Geoprocessing > Union.
120. Select Union_Soil_PPT as the Input layer.
121. Select Arequipa_Slope_Polygon as the Overlay Layer.
122. Save the output in your working directory as Union_Soil_PPT_Slope.
123. Click Run.
124. Open the attribute table.
125. Scroll to the right, and we should see the three newly added columns, Soil_Type_, PPT, and slope.

We now create a new attribute that contains information for these three layers.

126. Open the Field Calculator.
127. Create a new field called 'Unique'.
128. Set type to Text.

We want to combine the values in each attribute, and we do this through a function called concatenate.

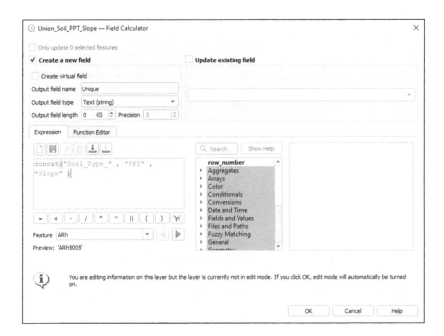

FIGURE 16.15 Screenshot of the concat expression in field calculator.

129. Type the expression: concat ("Soil_Type_", "PPT", "Slope").
130. Ensure your dialogue box resembles Figure 16.15. Click OK.

In the attribute table, there is now a new attribute with each feature having its own unique combination of soil, slope, and ppt, e.g., ARh5008 or LPe5005, shown in Figure 16.16.

We can repeat the spatial join to identify unique combinations at the location of the landslides, but first we must save any edits that have been undertaken through Field Calculator.

131. Save Edits, either through the button in the Attribute Table or through the button on the Digitizing Toolbar. It is also useful to turn Toggle Edits off at this point.
132. Join attributes by spatial location, using the original Arequipa_Landslides file as the Base Layer, Union_Soil_PPT_Slope as the Join Layer, and selecting Unique as the attribute to return. Save this as Arequipa_Landslides_Unique in the GeoPackage.
133. Navigate through the tab Vector > Analysis Tools > List Unique Values. Save the table as UniqueTable.

This should return five unique conditions across the three variables, shown in Figure 16.17.

134. Navigate to Select by Expression, ensuring that Union_Soil_PPT_Slope is activated in the Layers Panel.

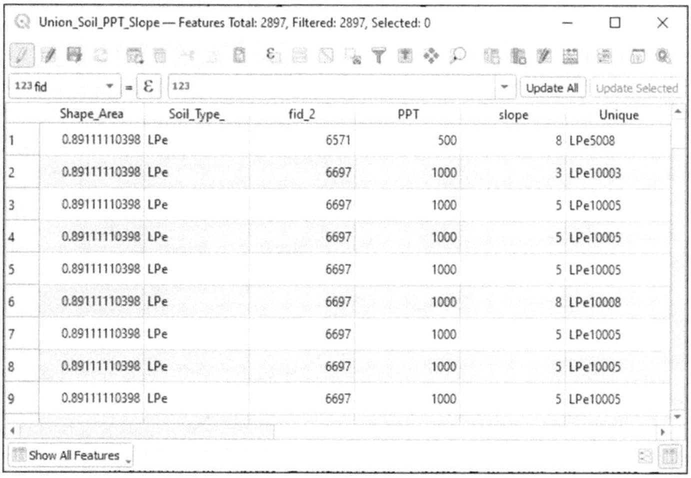

Risk Analysis – Unique Condition Unit

FIGURE 16.16 Screenshot of the attribute table showing the newly created unique variable.

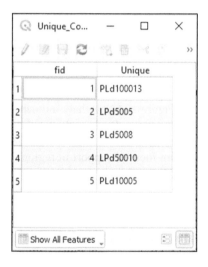

FIGURE 16.17 Screenshot of the attribute table showing the unique values of the combined soil-ppt-slope variable (unique).

135. Type the following expression:

"Unique" = 'LPd50010' OR "Unique" = 'PLd100013' OR "Unique" = 'LPd5005' OR "Unique" = 'PLd5008' OR "Unique" = 'PLd10005'.

136. Click Select Features (there are 365 features selected).
137. Save the resultant layer by right clicking on the Union_Soil_PPT_Slope > Extract > Save Selected Features As….

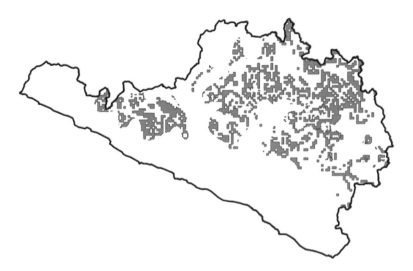

FIGURE 16.18 Results of the risk analysis that identify areas of landslide susceptibility in Arequipa, Peru, based on the unique conditions across all three environmental criteria.

138. Save this layer as Susceptibility_Refined in the GeoPackage.
139. Finally, for presentation purposes, dissolve the Susceptibility_Refined layer.

Our refined susceptibility map is shown in Figure 16.18. If we compare this to the output generated earlier in Figure 16.14, we have a much smaller area that could be considered susceptible to landslides. By considering the unique combinations, we can refine our mitigation strategies, inferring that all three combinations must be met. This could provide more focused support in regions that have a higher risk of landslides, thus supporting SDG11.5.

16.3 CASE STUDY CONCLUDING REMARKS

As mentioned earlier, caveats to this type of research include missing data, different variables, and real-time precipitation, but the workflow should reinforce the learning outcomes related to clip, intersect, and union, as well as spatial joins and more advanced field calculator options. It also flips the question from the previous chapter and rather than looking for the most suitable site we have been looking for the most at-risk site. These are perhaps the two most common types of questions that students undertake in GIS courses, but the majority focus on site suitability, despite the methods being broadly similar. This chapter has advanced our geoprocessing abilities and allowed us to combine raster and vector data seamlessly. In the next chapter, we continue to build on the ability of GIS to work with multiple layers; however, we use raster data, and specifically map algebra.

16.3.1 Test Yourself

If you would like to test yourself on the learning outcomes of this chapter, can you:

a. Use the skills from Chapter 4 to calculate the areal difference between the two areas of susceptibility?
b. Repeat this analysis for the whole of Peru?

REFERENCES

Akgun, A., Kıncal, C. and Pradhan, B., 2012. Application of remote sensing data and GIS for landslide risk assessment as an environmental threat to Izmir city (west Turkey). *Environmental Monitoring and Assessment*, 184(9), pp. 5453–5470.

Batjes, N.H., 2016. Harmonised soil property values for broad-scale modelling (WISE30sec) with estimates of global soil carbon stocks. *Geoderma*, 2016(269), pp. 61–68. DOI: 10.1016/j.geoderma.2016.01.034.

Becker, J.J., Sandwell, D.T., Smith, W.H.F., Braud, J., Binder, B., Depner, J.L., Fabre, D., Factor, J., Ingalls, S., Kim, S.H. and Ladner, R., 2009. Global bathymetry and elevation data at 30 arc seconds resolution: SRTM30_PLUS. Marine Geodesy, 32(4), pp. 355–371.

Goodchild, M.F., 2011. Scale in GIS: An overview. *Geomorphology*, 130(1–2), pp. 5–9.

Hijmans, R.J., Cameron, S.E., Parra, J.L., Jones, P.G. and Jarvis, A., 2005. Very high resolution interpolated climate surfaces for global land areas. *International Journal of Climatology: A Journal of the Royal Meteorological Society*, 25(15), pp. 1965–1978.

Huabin, W., Gangjun, L., Weiya, X. and Gonghui, W., 2005. GIS-based landslide hazard assessment: An overview. *Progress in Physical Geography*, 29(4), pp. 548–567.

Kirschbaum, D.B., Adler, R., Hong, Y., Hill, S. and Lerner-Lam, A. 2010. A global landslide catalog for hazard applications: Method, results, and limitations. *Natural Hazards*, 52(-3), 561–575. DOI: 10.1007/s11069-009-9401-4.

Kirschbaum, D.B., Stanley, T., Zhou, Y., 2015. Spatial and temporal analysis of a global landslide catalog. *Geomorphology*, 249, pp. 4–15. DOI: 10.1016/j.geomorph.2015.03.016.

Lee, S., 2019. Current and future status of GIS-based landslide susceptibility mapping: A literature review. *Korean Journal of Remote Sensing*, 35(1), pp. 179–193.

17 Site Selection – Map Algebra

17.1 INTRODUCTION AND LEARNING OUTCOMES

In this chapter, we build on the analytical methods learnt in the previous chapters, specifically slope, interpolation, proximity, and density by utilizing map algebra to undertake further site selection to support the SDGs. Unlike Chapters 15 and 16, in this chapter we use a raster workflow. The outcome is similar, whereby we use GIS to inform as to where best to locate a service; however, the data structure and methods implemented are different. If we think back through previous chapters, this very much echoes comments that there are multiple methods to achieve an output when working in GIS, and the differences between this chapter and the previous chapters in this section are perhaps a perfect example. Moreover, the methods introduced in Section 4 are not always an end product, but rather part of an intermediate process in the analytical framework.

Map algebra uses different operators to return numeric values to a new output. These operators can again be arithmetic, Boolean, or relational, and we implement such analysis by performing an expression between the raster layers in question. Consider that we have two spatial layers represented as raster datasets: precipitation and ground infiltration (shown in Figure 17.1). As we have a continuous representation of these variables, we can combine them to identify the amount of surface runoff we might expect across a study area. During precipitation events, there will be an amount of precipitation that will be infiltrated into the ground, while the rest will remain as surface runoff. Therefore, we can use map algebra to subtract infiltration from precipitation to capture the runoff. We can also employ such a method to capture site suitability or risk. Figure 17.1 also demonstrates an example of where physical and social variables are combined to provide a risk index of a natural hazard, in this example the risk posed by a tsunami. We have successfully used the raster calculator in Chapters 3 and 14, but in this chapter, it becomes our main analytical tool.

In this chapter, we use map algebra to identify a new site for an electric vehicle (EV) charging station in Vancouver, British Columbia, Canada. In doing so, we address SDG7.1 and SDG13.2. We know from Chapter 14 that SDG7.1 aims to ensure universal access to affordable, reliable, and modern energy services, while SDG13.2 aims to integrate climate change measures into national policies, strategies, and planning. A shift from petrol/diesel vehicles to EVs is a nuanced discussion, and while a national car fleet consisting predominantly of EVs is not necessarily the panacea for both SDG targets, it would certainly support sustainable transport and is an excellent case study to further site selection in a GIS context.

Vancouver has been selected as the case study for two reasons. Firstly, in 2020 they approved a Climate Emergency Action Plan, which included accelerating

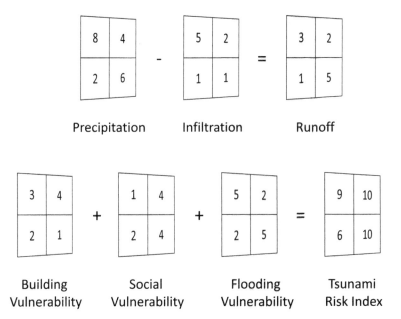

FIGURE 17.1 Conceptual diagram of map algebra to calculate surface runoff in the top row and tsunami risk in the bottom row.

the expansion of the public EV charging network, as well as adding incentives to accelerate the transition to EVs. As such, they are an excellent case study to explore how EV transition can be achieved. Secondly, the City of Vancouver has an extensive Open Data Portal (https://opendata.vancouver.ca/pages/home/), which at the time of writing has 177 datasets, of which most are spatially referenced. Subsequently, we can use this data to demonstrate how GIS can be used to support these SDGs.

The site selection process using raster layers works similarly to working with vector data, but obviously the functions and tools are different. The process consists of approximately four overarching steps. Step 1 is to source the spatial data. Step 2 is to pre-process this data, turning the raw data into information. For example, rather than location of existing EV stations, we are interested in the neighborhood or a specified distance that surrounds them. Step 3 is to reclassify these layers into binary or categorical outputs. The rationale here is that it is difficult to combine layers with different units (i.e., distance to EV chargers and slope angle). By categorizing these layers, it allows us to perform a simple map algebra expression. Finally, Step 4 is to combine and weight these layers through map algebra expressions. Figure 17.2 is a thematic sketch of the process we will undertake in this chapter of identifying a new suitable site for an EV charging station in Vancouver.

The variables used herein include roads, heritage sites, public streets, elevation, parks, and the local-area boundaries, all sourced from the City of Vancouver Open Data Portal (City of Vancouver 2021). All layers used in this chapter are licensed

Site Selection – Map Algebra 289

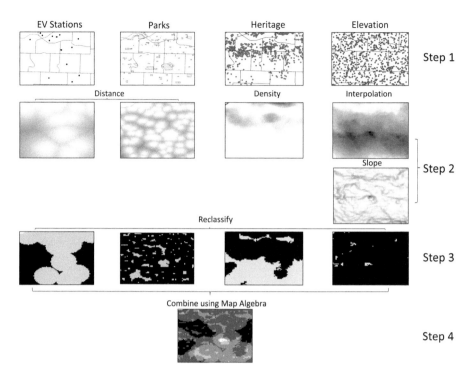

FIGURE 17.2 Conceptual workflow of the steps incorporated within the map algebra process for electric vehicle charging stations for an area within Vancouver, Canada

under the Open Government Licence – Vancouver. By the end of this chapter, you will have completed two learning outcomes, and should be able to:

- Perform several analytical tools to transform vector data to raster information, e.g., interpolation and density
- Weight and combine datasets using map algebra

17.2 CASE STUDY: LOCATING A NEW ELECTRIC VEHICLE CHARGING STATION IN VANCOUVER, BRITISH COLUMBIA, CANADA

1. Open a New Empty Project and add all the layers from the Ch17 Data.gpkg

We should notice that all the layers are currently in vector format, as shown in Figure 17.3. This is not unusual when working with GIS, especially as all these features represent discrete objects. However, in this chapter we want to explore how we can convert this data into spatial information stored in raster format.

We want to process the data using linear units (i.e., distance, density). This means we firstly need to check the projection of our layers.

2. Right click on the EV Charging Stations layer to open Properties, and identify the projection from the General Information tab.

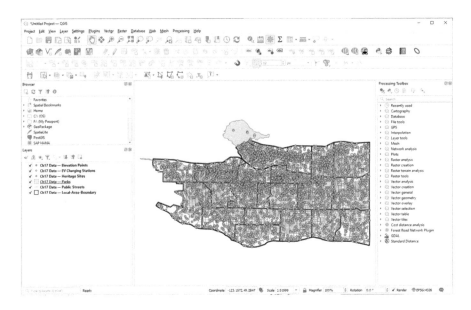

FIGURE 17.3   Screenshot of all the layers.

The layer is currently in WGS84, with no Cartesian coordinate system. We want to project to a system that will allow us to work in linear units, and following the standards set out throughout this book, we will use UTM. Vancouver is found in WGS84 UTM Zone 10N.

3. Therefore, the first task is to project all the layers into UTM Zone 10N, EPSG code 32610. Save these in your working directory, I suggest a new GeoPackage called 'Ch17_Projected'
4. Once this task is completed, we also need to change the projection of the project to EPSG 32610 in the bottom right of the QGIS interface.
5. Remove all the original layers to avoid confusion in the analysis. Hint, we can select all original layers in the Layers Panel using Shift and right click and remove all of them at once.

The layers could have been provided in the reprojected CRS, but it is important to acknowledge as we approach the end of this book, and subsequently the more quantitative analytical components of GIS, not to forget the fundamentals that were introduced in the earlier sections. This is especially pertinent when many of the functions and tools we use in this chapter (i.e., proximity, density) are impacted by the CRS. Therefore, the inclusion of these steps is to reinforce the learning outcomes from earlier chapters and consolidate GIS understanding.

There are two ways we could approach the overall site selection process in this chapter. Firstly, we could iteratively work through steps 1–4, converting all layers to rasters, then pre-processing all layers, before reclassifying and combining them in the map algebra process. Alternatively, we can complete the workflow for steps 1–3 for each individual layer, before combing the outputs in step 4. My preference, borne

Site Selection – Map Algebra

from experience and that of my students is the latter. Undertaking the process using this workflow allows us to engage with each individual layer and truly understand the whole analytical process, rather than repeating the same process for all layers, and potentially increasing the opportunity for mistakes or misunderstanding, which is increased when the workflow is not necessarily completed in one sitting (i.e., a two-hour break to attend another lecture). Therefore, the remainder of this chapter will work through the analysis by individual variables.

Finally, before we begin our analysis, we have implemented all the pre-processing steps in previous chapters, including proximity, density, interpolation, and slope. Therefore, now would be a good time to test yourself as to your newfound GIS abilities. Instructions are of course provided to ensure completion of this chapter but attempting to complete these steps independently will demonstrate completion of the learning outcomes associated with earlier chapters.

EV Charging Stations

The first variable we are interested in is existing public EV charging stations. When we are considering installing a new EV charging dock, we want to avoid locating this near any existing infrastructure to maximize the geographic impact of these stations by supporting more people. To do this, we need to identify a minimum distance within which we would ideally avoid. If we re-consider the concept of a 15-minute city that we explored in Chapter 11 (Moreno et al. 2021), we know that 1200 m (or 1.2 km) is the average distance that can be covered walking in 15 minutes. Therefore, to ensure a geographically widespread network of EV chargers, we want to ensure that any new ones are at least 1.2 km away from an existing network. The workflow up until the reclassification follows that of Chapter 11, whereby we rasterized a vector layer and used the proximity tool. I provide instructions below but attempt to replicate this analysis from your previous experiences.

6. Right click on the layer and open the properties
7. Navigate to the Fields tab.

There are three attributes: addresses, lot_operat, and geo_local_. To generate a raster representation, we need an integer field with which to point the rasterize tool to.

8. Click on Field calculator and create a new field EV as an integer. In Expression, simply type 1, which will set the values of all records in the point file to 1. Click OK, and then close the properties dialogue box
9. Open the attribute table to check this has correctly created the new attribute

These steps have opened up an editing session, which has not saved. To progress with our analysis, we need to save this.

10. Click on Current Edits, click save for selected layers. End the Toggle Editing session as well
11. Navigate through the tab Raster > Conversion > Rasterize
12. Set the input layer as EV Charging Stations
13. Set the Field to use for a burn-in as the new EV attribute

14. Change the output raster size to 'Georeferenced Units'
15. Set the width and height of the resolution to 100
16. Choose the output extent to match the layer of local-area-boundary
17. Save the output to your working directory using EV_raster
18. Click Run

Turn the other layers off to see the new layer. Currently, only the locations with 1 are returned by the raster. This is fine, as once we undertake some pre-processing, this will be completed at the extent we specify.

19. Navigate through the tab Raster > Analysis > Proximity (raster distance)

Again, we have used this tool before in Chapter 11, and we know it creates a value for each cell/grid that represents the Euclidean distance to the feature of interest, in our case EV charging stations.

20. Set the input layer as EV_Raster
21. Change Distance Units to Georeferenced Coordinates
22. Save the output as EV_Distance
23. Click Run

We should now have a raster layer that represents distance from each EV station within the geographic area.

24. Open the Processing Toolbox, and search for 'Reclassify by Table' and open the tool
25. Set EV_Distance as the Raster Layer
26. Save the output to your working directory as EV_Binary
27. Under reclassification table, click on the …

Here we specify the minimum and maximum values. Anything less than 1200 must be changed to 0 and anything above 1200 as 1. We therefore create two rows, using the specification that min < value <= max.

28. Complete the reclassification table as shown in Figure 17.4.

The table in Figure 17.4 is basically specifying the if else statement. If values are greater than 0 but less than or equal to 1200, they will have a value of 0, and if they are greater than 1200 (and for our purposes less than 8000), they will have a value of 1. Remember, areas which we are considering suitable should have a binary value of 1, and areas that we are considering unsuitable should have a binary value of 0.

29. Click OK. Remember our table needs to be closed otherwise this will bug out.
30. Click Run

We now have a raster layer that is represented using 0 and 1 for locations within 1.2km of EV stations, as shown in Figure 17.5. Zero represents locations that are within 1.2km and one represents locations that are beyond 1.2km.

Site Selection – Map Algebra

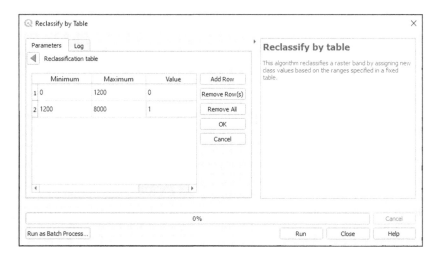

FIGURE 17.4 Screenshot of reclassification table.

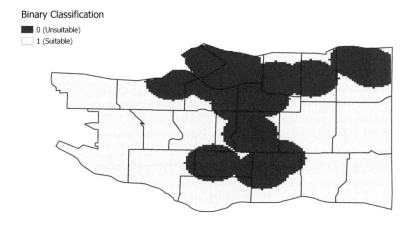

FIGURE 17.5 Screenshot of the binary layer for electric vehicle charging stations.

Before we progress, we should clip/extract this to the local-area-boundary to ensure that only locations in the city of Vancouver are returned.

31. Navigate through the tab Raster > Extraction > Clip Raster by Mask Layer
32. Select EV_Binary as the input
33. Select local-area-boundary as the mask layer
34. Select EPSG code 32610 as source and target CRS
35. Make sure 'Match the extent of the clipped raster to the extent of the masked layer' is ticked
36. Change the X and Y resolutions to 100
37. Save the output in your working directory as EV_Bin_Clip
38. Click Run

This is the layer that we integrate into the final step (step 4), so make sure this layer remains in the project. If the naming protocol has not been followed, make sure the name of this layer is noted as it will be needed later. We will be generating a lot of raster layers through this processing, and it will be easy to get lost if we do not keep note of the names.

Parks

The next variable we are interested in is parks. When we are considering installing a new EV charging dock, we want to locate this in places which the public will use. Therefore, locating a new EV station within a specific distance of an existing park will provide the infrastructure for people to charge their vehicle while undergoing recreation. Instead of 1.2km, we consider a suitable distance that is half this, 600m. This is close enough to the facility, without being so far that people may not want to walk to the park. It will also facilitate accessibility for those who might not be able to comfortably walk for 15 minutes. Again, step-by-step instructions are provided below to complete this section, but try to consolidate understanding by implementing this based on the knowledge acquired in the previous section of this chapter.

39. Open the properties of Parks
40. Navigate to Field and use the field calculator to create a new field called Parks. Set all the values to 1 using the expression option. Remember to save the edits and close Toggle Editing after completion.
41. Next, rasterize the parks polygon layer. Set the park polygon layer as the input, set the burn-in value to the newly created Parks attribute, set the output raster size to 100 georeferenced units, and set the extent as the local-area-boundary.
42. Next calculate proximity on this Park_Raster layer, using georeferenced units
43. Reclassify this layer. Remember, reclassifying all locations between 0 and 600 as 1 (as these are the desirable locations) and beyond 600 as 0. We can put a max number of 5000, which should cover all distances greater than 600.
44. Finally, use the clip raster by mask layer tool to clip the reclassified raster of distance to parks to the extent of the local-area-boundary.

We now have a raster layer that is represented using 0 and 1 for locations within 600m of a park, as shown in Figure 17.6. Zero represents locations that are beyond 600m of a park, while one represents locations that are within that distance.

Heritage Sites

The next variable we are interested in is heritage sites. A lot of initial infrastructure for EVs is established for the resident population, meaning that facilities for visiting tourists are less prevalent. Therefore, when we consider installing a new EV charging dock, we want to locate this in a place which both the public and tourists may visit. Therefore, locating a new EV station within a high density of heritage sites will provide the infrastructure for people, specifically tourists, to charge their vehicle while exploring the built heritage. We again focus on a 1.2km distance to capture our metrics.

Site Selection – Map Algebra

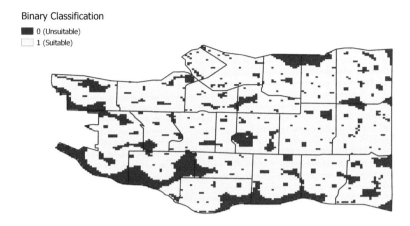

FIGURE 17.6 Screenshot of the binary layer for parks.

For this analysis, we use a kernel density to generate a heatmap, as we did in Chapter 13. The following steps will generate a density output for heritage sites, but note the additional steps from 52 to 56 that are newly introduced in this chapter.

45. In the Processing Toolbox, search for Heatmap (Kernel Density Estimation)
46. Specify the point layer as Heritage Sites
47. Set the radius to 600m. As this is a radius, we use half the distance of the 1200m
48. Change the pixel size to 100
49. Change the kernel shape to Uniform.
50. Save the layer to your working directory using the name Heritage_Density
51. Click Run

This creates a raster layer that represents the density of heritage sites with values ranging from 1 to 316. We should also notice that this layer does not cover the entirety of the local-area-boundary, with locations beyond 600m of any heritage site given a NoData value. We have discussed the importance of NoData in previous chapters; however, unlike Chapter 13, we can be confident that a NoData value within the study area is actually equivalent to zero (i.e., there are zero heritage sites within that area). This is important, as when we come to the final stage in this analysis, we overlay the binary raster layers of the four variables; however, if there are not the same extent, QGIS will automatically clip the processing to the smallest extent. This means large parts of our study area will be ignored. Therefore, we want to change the values of NoData to 0.

52. Navigate to the Processing Toolbox and search for Fill NoData Cells. This should be under Raster tools. Open it
53. Specify the raster input as Heritage_Density
54. Change the fill value to 0
55. And save the output raster to your working directory as Heritage_Density_Complete

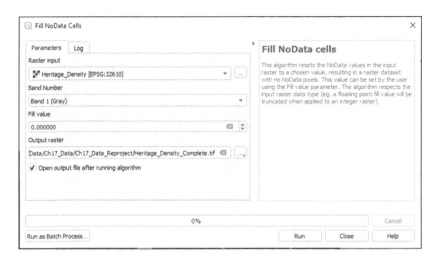

FIGURE 17.7   Screenshot of the parameters for the Fill NoData Cells tool.

56. Ensure your dialogue box resembles Figure 17.7. Click Run

We should now see that the NoData values have extended to the overall extent, and largely match those that we have been generating thus far (although not perfectly – but more on that in a moment). Now we want to classify these to represent binary values.

57. Open the reclassify by table tool, specify the Heritage_Density_Complete as the input layer

We want to reclassify high density of heritage points to 1. For this example, we consider a density of 10 heritage sites within the 1.2km neighborhood. This is purely an arbitrary number, but probably a suitable number of sites for tourists.

58. Therefore, in the reclass table set the first row (min 0, max 10, value 0) and the second row (min 10, max 320, value 1)
59. Keep the Range boundaries as default
60. Click OK, and save the output as Heritage_Binary
61. Click Run
62. Finally, use the clip raster by mask layer tool to clip the reclassified density of heritage sites to the extent of the local-area-boundary.

We now have a raster layer that is represented using 0 and 1 for high heritage density within a 1200m neighborhood, as shown in Figure 17.8. Zero represents locations that have a density of heritage sites of less than 10 per 1.2km, while one represents locations that have a density higher than 10 per 1.2km.

Elevation
The final variable we are interested in is elevation. The development of EV infrastructure will be easier and cheaper in flat areas, as well as assuming that the

Site Selection – Map Algebra

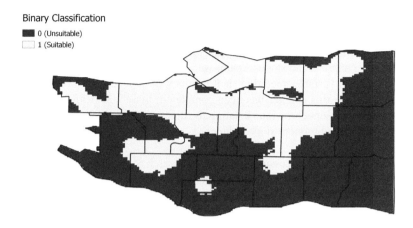

FIGURE 17.8 Screenshot of the binary layer for heritage sites.

public/tourists will most likely park in flat areas if they are then subsequently exploring the city. The city of Vancouver is relatively flat. Although I can recount an outing a few years ago where the travel guide for the Grouse Grind in north Vancouver was not read in enough detail, resulting in a 3km vertical climb up the mountain in non-walking attire… However, we include elevation as an input variable as it provides an opportunity to reinforce learning outcomes associated with interpolation and slope.

The City of Vancouver provides 1m elevation data on their open portal, but the size of this dataset when zipped is approximately 1.5GB. Because of this, the data used herein is a 2000 point sample of this elevational dataset, meaning we must interpolate it into a raster layer. Obviously, there is no real-world reason to convert a raster to points before interpolating it back to a raster, other than to support GIS learning outcomes.

Again, building on the skills from Chapters 14 and 16, try to interpolate the points and create a slope. Use an IDW interpolation method with a distance coefficient p of 3, with the analysis completed to the extent of the local-area-boundary. When completing the slope analysis, the z factor can remain as 1 as all X, Y, and Z values are metric.

63. Navigate to the IDW Interpolation tool
64. Specify ElevationPoints as the Vector Layer
65. Specify RASTERVALU as the interpolation attribute
66. Click the green +
67. Set the distance coefficient p to 3
68. Set the Extent to match the local-area-boundary layer
69. Change the pixel size to 100 to match the other raster layers
70. Save the layer as Elevation in your working directory
71. Click Run

The elevation data was originally captured at 1m resolution, meaning the data has undergone a resampling to 100m, which is quite a coarsening of the data. Therefore, if we were to implement the interpolation using a finer resolution, we would have an interpolation that accounts for more variability in the topography. However, for the purposes within this chapter, and to demonstrate the learning outcomes, it is sufficient.

Next, we calculate the slope. Again, we should pause here to note that slope is a scale dependent variable, meaning such results will depend on the resolution of the input data. Once this chapter has been completed, these steps can be revisited and investigated to observe how interpolating the data to a finer resolution (1m, 5m, 10m, etc) changes the subsequent slope values.

72. Navigate to the Processing Toolbox, and search Slope. This should be in the Raster Terrain Analysis toolbar. Open the tool
73. Select the Elevation as the input layer
74. Keep the z factor as 1
75. Save the output to your working directory as Slope
76. Click Run

The slope values range from 0.01° to 12.5°. Flatter surfaces provide support when building infrastructure, so we must reclassify this layer to identify locations that have a slope of less than 5°.

77. Open the Reclassify by Table tool
78. Set Slope as the input variable
79. Set the reclass table to the first row (min 0, max 5, value 1) and the second row (min 5, max 15, value 0)
80. Save the output in your working directory as Slope_Binary
81. Click Run
82. Finally, use the clip raster by mask layer tool to clip the reclassified slope layer to the extent of the local-area-boundary.

We now have a raster layer that is represented using 0 and 1 for flat surfaces, as shown in Figure 17.9. Zero represents areas that have a slope greater than 5°, while one represents flat areas that have a slope less than 5°.

Step Four – Combining and Weighting Layers

We now have four processed layers, including EV_Binary, Park_Binary, Heritage_Binary, and Slope_Binary. For ease of work, turn all layers other than these off or remove any additional ones we no longer need (with the exception of local-area-boundary). As we have created these layers using slightly different datasets, the extents are slightly different. Therefore, to implement an effective overlay using map algebra, it was imperative that we clipped them to the local-area-boundary to ensure that only locations in the city of Vancouver are returned. If you have not done so already, you will need to complete that step of the analysis workflow to progress.

We now have four binary raster layers representing suitable (1) and unsuitable (0) conditions to locate a new EV charging station. Next, we need to combine these using the raster calculator function. The logic here is that by summing the four layers, locations which satisfy all four of the criteria will have a value of 4, locations that satisfy three out of the four criteria will have a value of 3, and so on...

Site Selection – Map Algebra

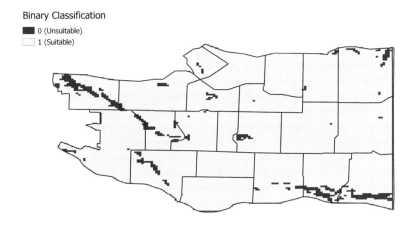

FIGURE 17.9 Screenshot of the binary layer for slope.

83. Navigate through the tab Raster > Raster Calculator

Here, we want to build the following expression:

"EV_Bin_Clip@1" + "Heritage_Bin_Clip@1" + "Park_Bin_Clip@1" + "Slope_Bin_Clip@1"

84. Double click on EV_Bin_Clip@1 to return it to the expression box, then click on the plus sign. Repeat this for the four layers.

Remember the @1 specifies the band number, which is still needed even if there is only one band.

85. Save the layer in your working directory as SuitableSites
86. Click OK
87. Right click on the new layer, navigate to the Symbology tab in Properties
88. Symbolize the layers using unique values with a blue color palette. Click classify. The dark blue represents colors that satisfy the 4 criteria.

There are approximately 5–7 geographic areas within the city that meet all four criteria, as shown in Figure 17.10. These are all primarily in the northern and western parts of the city. There are also some locations that match none of the criteria in the center of the city. Therefore, if tasked with this project, we could conclude that the ideal location would be to locate a new EV charging station/dock in one of these areas.

Here we have overlaid these four layers to identify locations that satisfy all four criteria. Luckily, we have locations that match all four criteria, but in reality we may wish to include several more variables, with the likelihood of finding locations that match all of the criteria increasingly unlikely as that number increases. This is where we could weight features/variables with a higher ranking. If we repeat this analysis

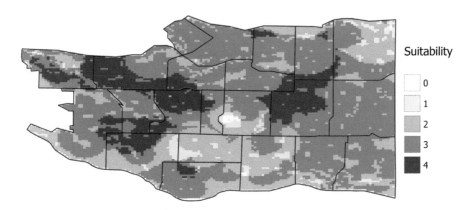

FIGURE 17.10 Result of the map algebra site suitability for a new electric vehicle charging stations.

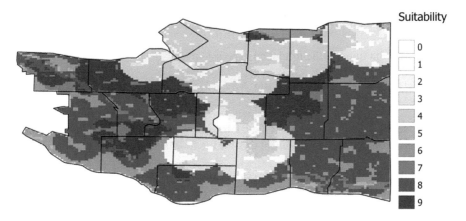

FIGURE 17.11 Results of weighted map algebra site suitability for a new electric vehicle charging stations.

weighting avoidance of existing EV stations as 5, closeness to parks as 2, and keep the rest as 1, we may get different results. Because our layers are binary (1,0), when we multiply EV_Bin_Clip, we subsequently have values of 5 and 0. When we sum these to the remaining layers, they will carry five times the weight.

89. Reopen the raster calculator, and type the following expression, remember to include the parentheses/brackets:

("EV_Bin_Clip@1"*5) + ("Park_Bin_Clip@1"*2) + "Slope_Bin_Clip@1" + "Her_Bin_Clip@1"

90. Save the layer as SuitableSitesWeight
91. Click OK
92. Change the symbology to blue

As we had locations that matched all 4 criteria, the best locations have not changed. However, we can see greater variability between the best and the worst locations on our new scale of 1 to 9. If we consider that we did not actually have any locations that matched all four criteria, and our highest location scored 3, when we apply weights, we may end up with a slightly restricted geographic location of what our model considers 'best'.

17.3 CASE STUDY CONCLUDING REMARKS

This chapter represents a culmination of many of the analytical tools that we have used throughout the book. If you completed the pre-processing of the variables without following the instructions, well done. If you did follow the instructions, it should have reinforced the skills that we have been working on previously, and result in a high level of competency of geoprocessing. The combination and weighting of these variables through map algebra provides another method of integrating spatial datasets and performing spatial analysis to address new questions.

While Vancouver City has made large strides for EV chargers, using our simple map algebra we have identified a relatively large geographic area where it would be highly suitable to develop more, especially supporting the two SDG targets outlined above. In Chapter 15, the importance of refining the suitable sites to one specific location was stressed. Given the large area returned in this analysis, this might not be simple and would most likely be achieved through the addition of more variables.

17.3.1 Test Yourself

If you want to test yourself on the learning outcomes of this chapter, complete the following:

a. Return to the City of Vancouver Open Data portal and identify a further variable that might be worthwhile including in the map algebra analysis?
b. Restrict the study area further to locations that are on residential streets, using the public streets layer in Ch17 Data.gpkg? Hint you may need to use select by values and buffer tools to ensure that this layer can be used as a mask

REFERENCE

Moreno, C., Allam, Z., Chabaud, D., Gall, C. and Pratlong, F., 2021. Introducing the "15-Minute City": Sustainability, resilience and place identity in future post-pandemic cities. *Smart Cities*, 4(1), pp. 93–111.

18 Route Selection

18.1 INTRODUCTION AND LEARNING OUTCOMES

In this final case study chapter, we explore how GIS can be used to identify route selection. The question we are essentially asking is 'what is the best route to take?'. This question has become synonymous with everyday life, as many of us have experience using navigation apps on our smartphones to navigate through cities when walking or driving. When we use these apps, our devices undertake spatial analysis to identify the optimal route, selecting the shortest or quickest route or those that avoid certain features, such as toll roads. However, such analysis could easily be extended to consider 'other' features that might be important to people's mobility choices.

There is a large amount of research that has been undertaken investigating why people select the routes they do. There is a growing body of literature that is recognizing that for cities to be truly sustainable, they need to be inclusive of different spatial perceptions, particularly when safety is considered (Ji et al. 2021). For example, Coakley (2003) explored mobilities of women in Cork City, Ireland, identifying that when moving around the city at night, most individuals opted to use main roads in well-lit areas, avoiding areas known for crime, such as parking lots, railway lines, and abandoned buildings. Similarly, Rashid et al. (2019) found that 75% of women surveyed agreed that presence of CCTV made them feel safer while moving through urban environs in Kuala Lumpur, Malaysia. Rosenberg et al. (2013) surveyed older adults with mobility disabilities in Washington, USA, with key themes related to route selection identified including curb ramp availability, lighting, presence of crosswalks, and safety, among others. These types of geographic features could be incorporated into the route selection algorithms we use in GIS, particularly to support the SDGs.

In this chapter we explore how GIS can be used to inform route selection specifically addressing SDG11.2. This target aims to provide access to safe, affordable, and sustainable transport systems, with specific attention to the needs of those in vulnerable situations, including women, children, persons with disabilities, and older persons. Such work could also support SDG5b, which aims to enhance the use of enabling technology to support and empower women.

By the end of this chapter, you will have completed three learning outcomes, and you should be able to:

- Conceptualize movement across both raster and vector data models
- Implement least-cost paths and shortest route algorithms
- Develop an environmentally weighted movement resistance layer

18.2 CASE STUDY: SDG11.2 WHAT IS THE BEST ROUTE TO TAKE TO THE TRAIN STATION?

1. Open the Ch18 QGIS project file.

In this project file there are the following layers: CCTV locations (CCTV), derelict and vacant buildings (Buildings), roads (Roads), and an administrative boundary (CityBoundary). There should also be two other point layers, one representing the location of the train station and the other representing four possible tourist locations in the city. These are shown in Figure 18.1.

The Vacant and Derelict Land Survey (Buildings), Public CCTV locations (CCTV), and the Edinburgh Ward Boundaries (CityBoundary) were all sourced from the City of Edinburgh Open Data Portal (https://data.edinburghcouncil-maps.info/). These datasets are licensed under the Open Government License v3.0. These datasets are Copyright City of Edinburgh Council, contains Ordnance Survey data © Crown copyright and database right (2022). The Roads layer is the HOTOSM UK Scotland Roads dataset, sourced from the Humanitarian Data Exchange Portal licensed under the Open Database License (ODC-ODbL), and clipped to the Edinburgh area. The train station and tourist sites were digitized by the author.

To use route selection, firstly, we undertake a simple network analysis to identify the shortest route along the road network to the train station from all the tourist locations, the same method that might be applied if we were using our mobile device to navigate the city.

2. In the Processing Toolbox, navigate to Network analysis. Open the options.

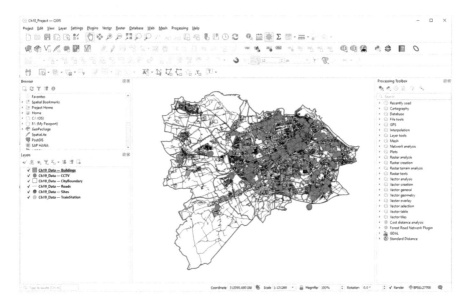

FIGURE 18.1 Screenshot of spatial layers in Edinburgh.

Route Selection 305

There are three options to calculate the shortest route, layer to point, point to layer, or point to point. Point to point is perhaps the simplest concept, as it will provide the shortest path between a and b. However, we are interested in multiple locations that are stored in a layer, meaning we must select one of the methods with a layer to point. It does not really matter which method is chosen, as the results will ultimately be the same. For now, we are interested in moving from the tourist locations to the train station.

3. Double click on Shortest path (layer to point).
4. Select Roads as the Vector layer representing the network.
5. Choose Shortest as the path type to calculate.
6. Choose Sites as the vector layer with the start points.
7. Click on the ... for end point.

This will generate a bullseye marker that will select the exact location of the end point. This is the point that represents the train station.

8. Zoom into the train station and click on the point.

This returns us to the dialogue box. There are various functions within this tool that we could change here. The main aim of this chapter is to demonstrate how to link movement with the underlying environment, so to ensure no methodological artifacts, we keep the default direction as both directions. There are also options for speed and direction. If we were driving, this would be relevant, but for now it is not. We return to this later in the chapter.

9. Save the file in the GeoPackage called Shortest_Distance.
10. Ensure your dialogue box resembles Figure 18.2. Click Run.

Examine how this identifies the shortest path from the various sites. Your results should resemble Figure 18.3.

This information provides us with the shortest route in geographic distance, but as outlined above, oftentimes routes are selected based on different factors, such as perceptions of safety. Again, we are feeding into SDGs 16 and 5, and considering how people perceive space and safety. There are a multitude of factors that people consider when selecting a route that is often individual and difficult to generalize, and everyone will have their own factors to consider when moving. While this has been relevant throughout the book, it's perhaps more pertinent in this chapter where we are discussing safety. For the purposes of demonstration, we generate routes in this chapter that have a higher or lower density of CCTV and derelict and vacant buildings, which have been found to inform decisions made on route selection for certain groups of people. This is done while acknowledging the fact that every individual will have their own perceptions of safe space, and that by simply perceiving a space as safe does not necessarily mean that it is true.

To demonstrate the most prevalent way this research is conducted, we need to install a plugin.

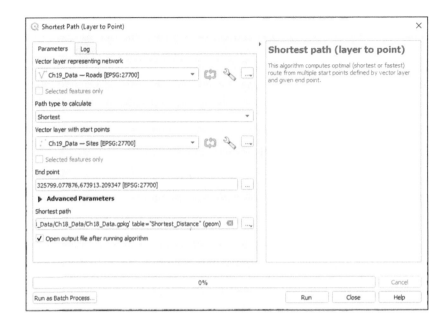

FIGURE 18.2   Screenshot of parameters needed for shortest path tool.

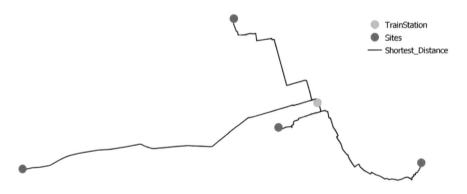

FIGURE 18.3   Results of shortest path tool.

11. Install the Plugin Least-Cost Path.

Least-cost paths are a route selection strategy based on a raster surface. The raster surface represents a 'cost' to movement, such as elevation or slope. The algorithm considers all options that are available and chooses the cheapest route relative to the cost surface, or in other words the path of least cost. It has been used widely in archaeology to identify historical roads (Lewis 2021), flood risk to identify the most likely route of water flow (Glas et al. 2020), and in ecology to identify migration paths and connectivity (Etherington 2016). Here, we use it in social context to identify the path of least resistance regarding safety.

Route Selection

We develop our own 'resistance layer' that considers two safety factors, the density of CCTV and derelict and vacant buildings. Density maps were explained in detail in Chapter 13, so for more information you can reference that chapter.

12. Navigate to the Heatmap (Kernel Density Estimation) tool in the Processing Toolbox.
13. Set CCTV as point layer.
14. Set the radius as 1200 (keeping consistent with our distance of 'walking city').
15. Set the raster pixel size to 100.
16. For these purposes we can keep the rest of the options as default.
17. Save the output in your working directory as CCTV_Den.
18. Click Run.

Again, there is no reason to assume that the CCTV layer is not complete coverage of CCTV cameras in the area, meaning we can safely assume that NoData is 0.

19. Navigate to Fill NoData Cells in the processing toolbox.
20. Set CCTV_Den as the Raster input.
21. Set the fill value to 0.
22. Save the output raster in your working directory as CCTV_Den_City.
23. Click Run.

Now we need to repeat these steps for the vacant and derelict buildings (buildings layer). However, this is currently a polygon, and we cannot capture density on a polygon layer, meaning we first must convert it to a point layer.

24. Navigate through the tab Vector > Research Tools > Random Points Inside Polygons.

This tool generates random points within polygons, meaning we can convert our polygon data to points.

25. Set Buildings as the input polygon layer.
26. Set the number of points in each feature to 1 (this will generate one point per polygon).
27. Save the output in the GeoPackage as BuildingPoints.
28. Ensure your dialogue box resembles Figure 18.4. Click Run.

Next, we need to repeat the previous analysis, calculating the heatmap for the building points and then filling in the NoData.

29. Repeat the steps to calculate density and change the NoData values to 0 with BuildingPoints as your input.

The raster outputs are different extents, shown in Figure 18.5. This is not an issue in this example, as the two extents cover the area for which we are interested in;

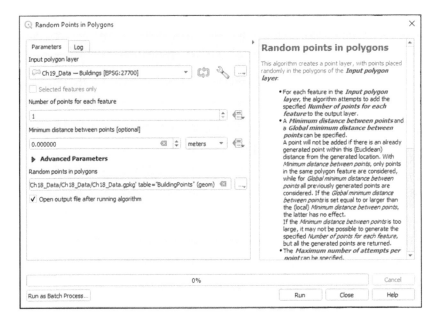

FIGURE 18.4 Screenshot of the parameters needed for the random points in polygons tool.

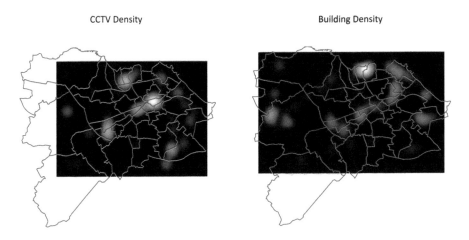

FIGURE 18.5 Density maps of CCTV and derelict and vacant buildings. Due to the interpolation method, the extents of the raster output are different.

however, we covered this in detail in the previous chapter should further explanation or revision be required.

Finally, we want to rescale our rasters so that the values are normalized. Therefore, we can use QGIS to set the minimum value as 0 and the maximum value as 100.

30. Navigate to the rescale raster in the Processing Toolbox in Raster analysis tools.

Route Selection

FIGURE 18.6   Screenshot of parameters for the rescale raster tool.

31. Set Building_Den_City as the input raster.
32. Set the new minimum value as 1.
33. Set the new maximum value as 100.
34. Save the output file in your working directory as Building_100.
35. Ensure your dialogue box resembles Figure 18.6. Click Run.

The output should not look any different, but the legend in the Layers Panel should now be between 1 and 100. We have selected 1 as the minimum value for a reason, instead of 0. This becomes evident in the next stage of this work when we link this back to the road network. Finally, we need to repeat this for CCTV; however, in this instance, a higher density of CCTV cameras is favorable. Therefore, when we rescale the raster, we invert it.

36. Open the Rescale raster tool.
37. Set CCTV_Den_City as the input raster.
38. Set the new minimum value to 100.
39. Set the new maximum value to 1.
40. Save the output file in your working directory called CCTV_100.
41. Ensure your dialogue box resembles Figure 18.7. Click Run.

Before we progress, we want to extract these layers to the boundary of Edinburgh. This will prevent our analysis going beyond the study area and into the North Sea.

42. Navigate through the tab Raster > Extraction > Clip Raster by Mask Layer.

FIGURE 18.7 Screenshot of parameters for the inverted rescale raster tool.

43. Specify the CCTV_100 as the Input Layer.
44. Specify the CityBoundary as the mask layer.
45. Specify the CRS as British National Grid.
46. Save the output as CCTV_100_City.
47. Click Run.
48. Repeat this analysis for Buildings.

We now have two raster layers on a scale of 1–100. The lower values mean ideal settings (high CCTV density, low derelict building density), which will equate to the path of least resistance in our models.

49. Navigate to the Least-Cost Path tool in the Processing Toolbox. This should have been installed with the plugin.
50. Choose CCTV_100_City as the cost raster layer.
51. Set Band 1 as the cost raster band.
52. Choose TrainStation as the start-point layer.
53. Choose Sites as the end-point(s) layer.
54. Save the output in the GeoPackage as a layer called Shortest_LCP_CCTV.
55. Ensure your dialogue box resembles Figure 18.8. Click Run.

We now have a new set of paths generated, as shown in Figure 18.8. We can see that the route is quite direct, perhaps too direct as it assumes that individuals can move in any direction over the raster and are not constrained by the road/path network. This is a limitation of the least-cost path method, particularly when working on linear

Route Selection 311

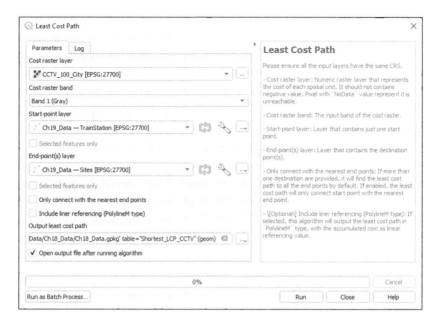

FIGURE 18.8 Screenshot of parameters for least-cost path tool.

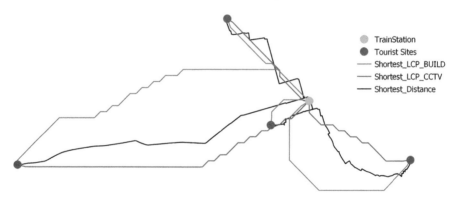

FIGURE 18.9 Results for the least-cost path route selection considering CCTV density (red) and building density (blue) compared to the shortest route along the network (black).

networks. There are workarounds to this, such as rasterizing the road network, but in the next section of this chapter we explore an alternative option. However, before we do this, let us repeat this process for the buildings layer.

56. Repeat these steps to calculate the least-cost path for buildings (Buildings_100_City).

The results are provided in Figure 18.9. This route subsequently avoids any location where there are derelict and vacant buildings within the density radius, taking the least-cost route, which in this instance takes a longer route to avoid certain parts of the city center.

If we turn the road layer back on, we can also observe that the paths do not correspond to the road network. Therefore, using such a conceptualization of movement and space may be unsuitable for this case study. Something we could consider here is that the least-cost path finds the shortest route based on this resistance, whereas if we were to consider the resistance as a speed, we could find the 'quickest' route to the location, not the least-cost. As mentioned earlier in the chapter, in the shortest path tool, there is the option to add a speed value.

Therefore, the next stage of this work is refitting the network shortest path tool but using our resistance values as a proxy for speed. Therefore, the assumption would be that more CCTV cameras or less buildings would allow for a 'quicker' route. However, to do this, we need to invert our resistance layers, because at the moment they represent cost, with higher values being undesirable.

57. Navigate to rescale raster.
58. Select CCTV_Den_City as the input layer.
59. Select Band 1 as the band.
60. Set the new minimum value to be 1.
61. Set the new maximum value to be 100.
62. Save the output in your working directory as CCTV_100_Inverted.
63. Click Run.

Because we are using this value to represent 'speed', we need a minimum value that is not 0. If the value is 0, it will assume that the individual will never be able to traverse the distance, and as such return no possible routes.

64. Navigate to rescale raster, selecting Building_Den_City as the input layer.
65. Select band 1 as the band.
66. Set the new minimum value to be 100.
67. Set the new maximum value to be 1.
68. Save the output in your working directory as Building_100_Inverted.
69. Click Run.

Next, we need to extract the raster values to the road network. If we think back to Chapter 3, we must use the two tools drape and extract Z values.

70. In the processing toolbox, navigate to Drape (set Z value from Raster).
71. Set Roads as the input layer.
72. Set CCTV_100_Inverted as the raster layer.
73. Save the output Draped file as CCTV_Drape.
74. Click Run.

This should have created a new road layer. Next, we want to summarize all the information that has been generated for each of the vertices.

Route Selection

75. Navigate to Extract Z values in the processing toolbox.
76. Set the Input layer as CCTV_Drape.
77. Click on the ... for summaries to calculate.
78. Ensure that Mean is the only option ticked. This will generate the mean 'resistance/speed' from the raster layer for the vertices of that line.
79. Save the column prefix as CCTV.
80. Save the output layer (Extracted) as Roads_CCTV.
81. Click Run.

Upon completion, we want to check that has worked.

82. Open the attribute table for Roads_CCTV.

If we scroll to the far right of the attribute table, we should see the mean raster value. Now, we can add the buildings values to this layer.

83. Navigate to Drape.
84. Set Roads_CCTV as the input layer.
85. Set Buildings_100_Inverted as the raster layer.
86. Save the output Draped file as Buildings_Drape.
87. Click Run.
88. Navigate to the Extract Z values.
89. Set the input layer as Buildings_Drape.
90. Set mean as the summaries.
91. Set the column prefix to BUILD.
92. Save the output layer as Roads_CCTV_BUILDS.
93. Click Run.

Again, check this has worked. Open the attribute table and ensure there are two columns, one for CCTV and one for Buildings. Now we can repeat the original shortest distance tool, but this time identifying the quickest route.

94. Navigate to Shortest Path (Layer to Point).
95. Set Roads_CCTV_Build as the vector layer representing the network.
96. Set the Path Type as the Fastest.
97. Set the start points as Sites.
98. Set the end point as the train station.
99. In advanced parameters, set the speed field as CCTV_mean.
100. Save the output as Shortest_CCTV_speed.
101. Click Run.
102. Again, before we explore the work, repeat this analysis with buildings as the speed variable.

The results are provided in Figure 18.10.

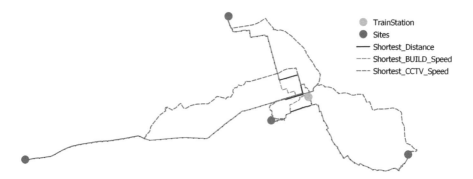

FIGURE 18.10 Results for the route selection considering CCTV density (red dash) and building density (blue dash) as speed values compared to the shortest route along the network (black).

18.3 CASE STUDY CONCLUDING REMARKS

In returning to the initial question of 'what is the best route to take?', we find that the answer is perhaps nuanced and certainly not always the 'shortest' or 'quickest' based on the road network. While these routes did not differ substantially in terms of distance, the choice of different routes based on geographic features could provide users with support in making navigation decisions in cities. This last case study chapter has identified some of the more innovative ways that GIS can be developed to support the SDGs, especially considering the applicability of this method to support app development while moving around cities.

The methods developed in this final chapter represent some of the topics that we touched upon in Chapter 10 related to dynamic entities. Therefore, it is perhaps a limitation of GIS that we are forced to represent mobility using a static spatial data model. Developing new methods that can account for space use within GIS, while acknowledging the dynamic processes that operate for objects that move, will be an area of growth within GIS over the coming years.

18.3.1 Test Yourself

If you want to test yourself on the learning outcomes of this chapter, complete the following:

a. Return to the Open Data portal and identify a further variable that might be worthwhile including in the route selection analysis, subsequently implementing it.

REFERENCES

Coakley, L., 2003. 'I don't relax until I'm home' women's fear of violent crime in public space in cork. *Irish Geography*, 36(2), pp. 178–193.

Etherington, T.R., 2016. Least-cost modelling and landscape ecology: Concepts, applications, and opportunities. *Current Landscape Ecology Reports*, 1(1), pp. 40–53.

Glas, H., De Maeyer, P., Merisier, S. and Deruyter, G., 2020. Development of a low-cost methodology for data acquisition and flood risk assessment in the floodplain of the river Moustiques in Haiti. *Journal of Flood Risk Management*, 13(2), p. e12608.

Ji, T., Chen, J.H., Wei, H.H. and Su, Y.C., 2021. Towards people-centric smart city development: Investigating the citizens' preferences and perceptions about smart-city services in Taiwan. *Sustainable Cities and Society*, 67, p. 102691.

Lewis, J., 2021. Probabilistic modelling for incorporating uncertainty in least cost path results: A postdictive roman road case study. *Journal of Archaeological Method and Theory*, 28(3), pp. 911–924.

Rashid, S., Wahab, M. and Rani, W., 2019. Designing safe street for women. *International Journal of Recent Technology and Engineering*, 8(2), pp. 118–122.

Rosenberg, D.E., Huang, D.L., Simonovich, S.D. and Belza, B., 2013. Outdoor built environment barriers and facilitators to activity among midlife and older adults with mobility disabilities. *The Gerontologist*, 53(2), pp. 268–279.

Epilogue

This book set out with the overarching aim to provide a practical-led approach to understanding GIS through the prism of the SDGs. If you have worked through this book, then you should now be competent in the use of QGIS for storing, managing, and analyzing spatial data. Many of the latter chapters provided the opportunity for you to implement your new GIS skillset in an independent manner, using functions and tools that did not necessarily require step-by-step instruction. The metaphor of removing the training wheels from GIS instruction is apt, as this process increases the opportunity for independent GIS work. This is an important part of the learning process, as to become proficient in the use of any GIS software we need to extend and develop our base skills to further advance our understanding.

Throughout this book we have used several different QGIS tools, functions, and plugins; however, as you may have noticed as we entered the final section of the book, there is an immense amount of functionality within this software that we did not cover. It is simply not feasible to write a book that covers every tool, nor in my opinion would it have been conducive to understanding GIS. We focused on the core tools and functions to explore real-world case studies related to the fundamental components of GIS, cartography, and spatial analysis including measurement and geoprocessing tools. The natural sixth section of this book in my opinion would be the demonstration of spatial statistics, such as spatial autocorrelation and non-stationarity.

For example, the plugin 'Hotspot Analysis' by Oxoli et al. (2017) provides the opportunity to implement local indicators of spatial autocorrelation (LISA). Spatial autocorrelation is grounded in Tobler's (1970) first law of geography and can be used to provide a statistical representation of clustering. In Chapter 5, we explored how the attribute table and queries could support SDG9c, universal access to broadband. LISA analysis has been used to a similar effect, with Gruebsic (2006) using areas of negative spatial autocorrelation (i.e., an area of high connectivity surrounded by areas of low connectivity) as support for policy intervention, terming these areas as 'islands of availability' as these represent geographic locations where the infrastructure for high connectivity was evidently established, meaning the cost of extending the infrastructure would be lower. This Plugin is dependent on Python Spatial Analysis Library (PySAL), and partly because of this dependency it is still experimental. As stated, to ensure the main aim of this book in supporting a comprehensive understanding of GIS was met, we did not explore the possibilities of experimental plugins. However, as QGIS continues to grow, the importance of the wider community and development of these plugins to support more complex spatial statistical analysis will become more pertinent.

Another Plugin that we could have used in Chapter 11 when investigating the spread of the Argentine ant was the GBIF Occurrence plugin. This links to the Global Biodiversity Information Facility (https://www.gbif.org/), which is an open data portal of biodiversity. There are over 2 billion occurrence records of animals,

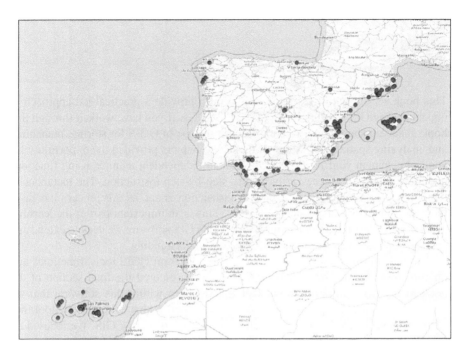

FIGURE 1 Locations of the Argentine ant in Spain loaded from GBIF Occurrence plugin. Basemap is the OpenStreetMap XYZ tiles which is © OpenStreetMap contributors and available under the Open Database License. Please see https://www.openstreetmap.org/copyright.

plants, and other taxa on this portal, and an excellent resource if working with biodiversity data. If we were to install this plugin, we would simply need to specify the scientific name of the species (*Linepithema humile*) and the country location (Spain) and click load occurrences. As we can see above, there are several occurrences for this species in Spain, but only two observations in Madrid in the center of the country. This dataset supports the overall pattern of this invasive species being predominantly found in coastal regions in Spain but requires more input to align with the detailed systematic surveys from López-Collar and Cabrero-Sanudo (2021) in Madrid. At the time of writing, there are 866 Plugins available in QGIS, with a further 231 experimental Plugins. Therefore, independently exploring the Plugins available in QGIS and the resources generated alongside these will substantially enhance our GIS work.

QGIS supports the use of the programming language Python, which can be used to extend the core functionality of the software, as well as writing bespoke script and tools, which among many other things can automate our workflow. There can be complications when working with Python across operating systems, as well as the somewhat steep learning curve associated with beginner programming. Python in QGIS or PyQGIS as it is referred to will continue to develop over the foreseeable future. Therefore, if you feel competent in your newfound QGIS practical skills and wish to develop these further through a programming language, there are resources available to support this. The use of programming languages to support GIS is continually growing and such a skill is increasingly being identified as desirable or even essential job criteria in our field, meaning even a basic understanding could increase your GIS functionality and support career development.

Compatibility with other GIS software is also a strength of QGIS. GRASS-GIS is another well-known open-source GIS software that was first developed in the 1980s. There are several functions and tools developed within this software that can be accessed through QGIS. Similarly, the Geospatial Data Abstraction Library (GDAL) can also be accessed through the QGIS interface, meaning users have access to the wide array of vector and raster tools developed within. As such, QGIS is perhaps leading the way in GIS software coexistence, as it means that users can benefit from years of GIS development across platforms. This book did not explore any of these tools for the reasons outlined above, but users are certainly encouraged to explore these options now that we are familiar with the core functionality of QGIS.

Finally, with new technologies come new challenges. Advances in the technologies we use to collect spatial data, as well as improved processing power that can support high-resolution spatial data-analytics means that we continue to ask increasingly unique and novel questions that can support many of the pressing environmental and societal questions we are currently facing. In Chapter 10, we explored how spatiotemporal data can be used to advance beyond static maps to represent a dynamic visualization of zebra migration. Similarly in Chapter 18, we considered how GIS could capture spatial experiences instead of simply observations of spatial data. Such frontiers in GIS are linked with the disaggregation of spatial data to explore the patterns and processes associated with individuals and networks, moving beyond the static to the dynamic.

Visual analytics is also an active analytical framework within GIS, so it is important to consider how cartography can be an important independent component for analyzing and modeling geographic phenomena. Cartographic representations as explored in this book are important for communicating results, generating knowledge, and validating our models. However, improving animation will be important research frontier for GIS, with the in-built functionality of QGIS supportive of such an endeavor.

As it becomes easier to model complex systems, the veracity of the models built will begin to be scrutinized. Many of the models explored within this book utilized case studies that touched upon a handful of variables to include when in reality, there will be several other considerations that must be met. Continuing to ground these models in domain-led theory and rationale will be important. In other words, just because we can model something one way, it does not necessarily mean we should. This was the rationale behind demonstrating the various outcomes in the latter chapters of this book, highlighting the importance to consider fitting multiple parameters in these GIS tools, and not simply selecting the default options. GIS is well placed to address these challenges and I hope you have enjoyed working through these case studies.

Finally, the real-world applicability of GIS to support practical decision making is perhaps one of its strongest advantages, and it was through the prism of the SDGs that we worked through global case studies that spanned multiple disciplines. As stated in Chapter 1, it was my hope that many of the methods, ideas, and analysis resonated with you, and that this book has allowed the possibility for you to consider GIS to support your own research and work.

Paul.

REFERENCES

Grubesic, T.H., 2006. A spatial taxonomy of broadband regions in the United States. *Information Economics and Policy*, 18(4), pp. 423–448.

López-Collar, D. and Cabrero-Sañudo, F.J., 2021. Update on the invasion status of the Argentine ant, Linepithema humile (Mayr, 1868), in Madrid, a large city in the interior of the Iberian Peninsula. *Journal of Hymenoptera Research*, 85, p. 161.

Oxoli, D., Prestifilippo, G., Bertocchi, D. and Zurbaràn, M., 2017. Enabling spatial autocorrelation mapping in QGIS: The hotspot analysis plugin. *GEAM. Geoingegneria Ambientale e Mineraria*, 151(2), pp. 45–50.

Tobler, W.R., 1970. A computer movie simulating urban growth in the Detroit region. *Economic Geography*, 46(sup1), pp. 234–240.

Index

add field 73–74
add geometry attributes 59–60
adjacency 69, 78, 177–178, 226–227
attribute table 7, 24–36, 51–60, 63–81, 86, 95, 98, 112–119, 134–135, 147–150, 166–174, 192–201, 206–210, 215–221, 223–228, 268–291, 313–317

basemap 15–17, 49–55, 77–78, 111–130, 164–173, 181–187, 208–234
basic statistics for fields 256–258
Boolean logic 69, 221, 259, 264, 287
browser panel 15–16, 36, 49, 63, 84–88, 111, 208, 224
buffer 79–81, 93, 94, 177–178, 184–187, 202, 252–254, 258–265, 301

cartographic elements: balance 7, 123–131, 177; legend 109, 113, 124–131, 142–143, 155–157, 188–190, 309; north arrow 109, 124–131, 142, 157; scale bar 109, 124–131, 142, 157
check validity 96–99, 256
clip 83, 148, 191–192, 200–202, 243–245, 251–254, 264, 267–269, 284, 293–300, 304, 309
coordinate reference systems (CRS) 49–61, 87, 91, 184, 191, 193, 200, 224, 243, 253, 257, 263, 290, 293, 310
create grid 193–194

data types: datetime 87, 166–168, 207; floating/decimal 39, 73, 148–149; integer 39, 65, 73, 100–102, 113, 134, 227, 278, 291
datum 43–50
difference 254–256, 261–264, 267
digitization 13, 78, 96, 178–183, 196, 202, 282
dissolve 184–187, 252, 254, 257–262, 280, 284
distance: Ellipsoid/Geodetic 57–59, 217; Euclidean 177–178, 292; Manhattan 177–178
dock 13, 27–28, 30, 51, 66, 121
drape 35–36, 312–313

Ellipsoid 45–46, 50–51, 56–60
EPSG (European Petroleum Survey Group) Codes 52–59, 80, 87, 91, 109, 180, 184, 193, 224, 253, 263, 290, 293
erase *see* difference

export map animation 172
extract Z values 35–37, 312–313

field calculator 63, 75–76, 101, 147–149, 194–195, 227, 234, 281–284, 291, 294
Fill NoData Cells 295–296, 307
filter 116–120
fix geometries 83, 98–99, 256

geoid 45–46
geopackage 83–95, 98, 101, 112–113, 158, 164, 180–184, 191–193, 196, 206–212, 215, 217, 229, 239, 252–261, 269–284, 290, 205, 307, 310
Global Positioning Satellite (GPS) 45, 163–164, 167, 173, 230
gray area 11–13, 15, 34, 57, 167, 180

intersection 148, 260–264, 267, 279–280
Inverse Distance Weighting (IDW) 240–247, 297
isochrone 186

join: by location 259, 269–284; table join 64–68, 231–135, 148

Kernel Density Estimation 230–231, 234, 295, 307

latitude & longitude 43–46, 50, 109, 142, 179–183, 210, 263, 275
layers panel 15, 25–26, 30–32, 49–52, 55, 64–69, 77, 89–96, 100, 111–116, 121, 180–181, 186–197, 208–212, 216, 231, 243–245, 257, 263, 272, 275, 278–282, 290, 309
least-cost paths 303, 306–312
list unique values 271–282
long-term release 9–10, 65

map canvas 11–15, 25–28, 56–58, 110–111, 122–127, 147, 157–159, 171–172, 182–184, 193
map types: change 107–108, 147–162; choropleth 107–108, 133–145; graduated symbols 107, 115–118; heatmap 108, 223–235, 240, 295, 307; inset 127–130, 147, 157–161; location 107–115, 130
mean center 205–221
measure line tool 56–59, 217, 264
metadata 17, 67, 85–86, 188

321

nodata 68, 74, 140–142, 194, 198, 209–210, 218–220, 231–232, 269, 276–277, 295–296, 307
null *see* nodata

package layers 87–88
plugins 5, 8, 12–13, 16–17, 121, 180, 183, 211–212, 305–306, 310, 317–318
polygonize 38–39, 255–256, 278
Print Composer 109, 122–131, 141–144, 155–161
Processing Toolbox 12–14, 30, 34–36, 87, 98–101, 113, 193, 197, 201, 211, 230, 240–242, 256, 270, 274–278, 292, 295, 298, 304, 307–313
proximity 177–178, 196–198, 202, 291–294

QML Style File 155, 189–190
QuickMapServices 16–17, 121–122, 165

random points in polygons 307–308
raster bands 30–33, 240, 255, 277–278, 299–301, 310, 312
raster calculator 32–35, 39, 245–247, 287, 298–300
rasterize 37–40, 193–196, 291–292, 294, 311
raster layer unique values report 201–202
reclassify by table 197–199, 277, 288, 290–298
refactor fields 101–102, 113
rendering 32, 99, 118–124, 151, 200, 225 226, 232, 234
reproject layer 54, 184–185, 191–192, 253–254, 290
rescale raster 308–312

Sample Raster Values 34–35
Save Selected Features As 51, 78–79, 90, 191, 208, 229, 257, 273, 275, 279, 283
save vector layer 53, 91
scale: extent 28–29, 38, 52, 64, 123–124, 141–142, 148, 157–158, 161, 171, 191–193, 196, 237, 240–243, 269, 273–274, 292–298, 307–308; map scale 29, 124, 226; resolution 37, 148, 174, 187, 193, 196, 201–202, 276, 292–293, 297–298, 319

select: by expression 207–208, 210, 220, 263–264, 272–273, 275, 279, 282–283; by location 78–80, 193–194; by value 69–73, 78, 89–90, 116–117, 191, 207, 229–230, 257, 301; interactive 28–29, 51–52, 55, 69, 76, 78, 112
shapefile 24–25, 39, 52, 83–87, 91, 101
shortest path 305–306, 312–313
slope 267–268, 275–277, 297–299
snapping 57, 217
spatial index 91, 99–100, 278
stacking 118–120
standard deviation ellipse 205–221
standard distance 205–221
sustainable development goals: SDG2 48; SDG3 4, 48, 205–206, 213–215, 217; SDG4 80–81, 133, 143; SDG5 223, 247, 303; SDG6 251–252; SDG7 238, 246, 287; SDG9 63, 70, 78–81; SDG11 163, 267–268, 284, 303–304; SDG13 23–24, 35, 40, 287; SDG14 108–109, 119, 128, 163; SDG15 48, 163–164, 174, 178, 187; SDG16 223, 228, 230
symbology tab 7, 25, 27, 32, 108, 113–121, 126–127, 133–142, 150–156, 159–162, 165–166, 183, 188–190, 200, 224–232, 270, 273, 299–300

temporal controller panel 167–172
temporal tab 166–167
temporary layers 92–96, 101–102, 201, 208
Toggle Editing 73–74, 77, 180, 196, 282, 291, 294
topology 79–81
Triangulated Irregular Network 239–241

union 251–252, 258–261, 264–265, 267, 281–284
unique identifier 58, 65–69, 81, 134–135
Universal Transverse Mercator (UTM) 50–51, 55, 60–61, 184, 191, 193, 252–254, 257, 290

vertice 23, 35–36, 77, 312–313

weighting 206, 210, 215–221, 227–229, 234–235, 264, 288–289, 298–301